A DENTAL ODYSSEY

[unlikely musings of a dentist]

by

Howard S. Selden

A Dental Odyssey
(unlikely musings of a dentist)
All rights reserved © 2009
by Howard S. Selden

A-Argus Better Book Publishers, LLC

For information:
A-Argus Better Book Publishers, LLC
Post Office Box 914
Kernersville, North Carolina 27285
www.a-argusbooks.com

ISBN: 978-0-9841342-7-4
ISBN: 0-9841342-7-1

Cover by DUBYA

Printed in the United States of America

Also by Howard S. Selden

Pariah Stigma
and
The Shaman and The Jew

DEDICATION

To Tamara—She's been there from
the beginning

CONTENTS

1.

DENTAL FIRST AID

Skiing at Taos, New Mexico, was memorable in an unanticipated way: I administered dental first aid.

~ * ~

Before I get into the dental story, a few words about Taos. It is a special area to which skiers are repeatedly drawn. After a long drive up a curving mountain road to over nine thousand feet, you arrive at the small ski village tucked in a high alpine valley among the towering Rocky Mountains. The feel is unforgettable: a Shangri-La-like place, out-of-this-world. The challenging ski slopes coupled with the charming, somewhat rustic, accommodations nestled unobtrusively among the trees, are a delight. And the remarkably good food alone is worth returning for.

But, significantly, there isn't a drug store. This deficiency challenged my dental ingenuity, as the following account will reveal.

That first ski day at Taos began as usual. We all lined up for a practice run down a gentle beginner's hill under the ski school's director's watchful eye for separation into small groups. The ski-off, as it is called, always made me nervous. If I skied poorly I was put in a humiliatingly low level group, while on the other hand, if by chance my technique appeared advanced, I might end up with experts—where I certainly didn't belong. Standing in the lift line deep in thought, I failed to notice that Dennis Beasley, the owner of the Over-the-Hill-Gang—the country's largest senior ski club and my host for this trip—was next to me. When I finally looked up, I was startled to find Dennis. His usual friendly smile was replaced by a grim expression. But what was really incongruous was the white rag hanging out of one side of his mouth.

Without hesitation, I knew he had a toothache. That portion of my brain assigned to dentistry clicked on automatically, even though I thought it was turned off when I retired a few years earlier. I am amazed how rapidly my mind sifts through alternatives when considering a differential diagnosis: often faster than it takes to talk about it. What helps is that the most likely diagnosis is often the correct one. Rare diseases are possible, but are indeed infrequent. If the initial impression proves incorrect, then the unusual must be considered.

I quickly discounted the possibility that Dennis's rag was used to control oral bleeding. Since we hadn't begun skiing yet, he couldn't have been hurt in a fall. Furthermore, such an accident would probably have also caused facial injuries; and there were none. The other possibility for oral bleeding could stem from the reopening of a recent oral surgical wound. In that unlikely situation, he would have surely been sufficiently alarmed to seek medical help at the first aid clinic.

Drawn out of my reverie I said, "Dennis, this is certainly your lucky day. I might be able to help with your toothache since I happen to be a dentist."

The rag almost fell out of his mouth as he lit up with surprise and delight. Quickly checking my

Over-the-Hill-Gang name badge—as a new member of the group we had never personally met—Dennis mumbled,

"Howard, how in the world did you know it was a toothache?"

"I could be wrong, but I figure you are using the rag to protect a sensitive tooth from the cold winter air. Tell me what's going on."

On our chair lift ride up the hill, Dennis told me his tale of woe.

"While eating a bagel for breakfast the plastic temporary cap on a back molar fell off. To my dismay the cap had split in half. Now whenever I inhale or even touch the tooth I have a painful jolt." He paused to reposition the rag, and through clenched teeth resumed talking. "My dentist recently worked on the tooth in preparation for a full gold crown, and until this accident the tooth was completely comfortable."

"I know I can help if you still have the temporary cap," I offered.

"Despite an initial reaction to throw the pieces in the garbage," he said, "I somehow resisted the impulse and dropped them in a glass."

"Swell, then," I said. "As soon as the ski-off is over, let's go back to the hotel before we start skiing and see what I can do."

Even with the rag in his mouth and feeling less than in a good mood, Dennis was assigned to the highest level ski group. I wasn't surprised since his reputation as an expert skier was widely known. I was relieved to be put in an intermediate group, hopefully not too gung-ho. Dennis explained to the director that we would be unavoidably delayed and would catch up to our groups in a little while.

~ * ~

The hotel was conveniently located across the bottom of the hill. A brief ski run brought us to the entrance where we planted our skis into the deep-piled snow, hung our pole straps over the ski tips, and clunked through the lobby in our bulky ski boots on our way to Dennis's room. After cleaning off the remnants of the temporary cement from within the sections of the plastic cap, I was pleased how well the two half's fit together. Next we went to my room for Krazy glue. I never leave home without a tube of this miraculous bonding agent (called cyanoacrylate) in my emergency kit. The glue didn't disappoint us. Within a few minutes the plastic cap was restored to full integrity, maybe stronger than ever.

Next I had to figure out what to use as temporary cement to seal the cap onto his molar. Without cement the relatively loose fit of the cap would allow cold air and liquids to painfully contact the tooth. Moreover, the loose cap could fall off, break, or even be swallowed.

I needed zinc oxide and eugenol to make a temporary cement but, without a drug store, these ingredients weren't available. This time-proven formulation is in routine use in dentistry to this very day. Mixing the eugenol liquid with the zinc oxide powder forms a sticky paste, suitable for holding temporary caps in place; preventing seepage of fluid or air from reaching the tooth. While the hardened cement is strong enough to hold the cap on the tooth, it is weak enough to allow the dentist to remove the cap. Eugenol is the active ingredient, creating a dense bond with the powder. Moreover, the eugenol has two useful effects: it is moderately germicidal; and can also ease and even stop toothache pain. Many sufferers learned from experience that a drop of eugenol into a painful cavity helps. The zinc oxide is pharmacologically inactive, serving as the bulk carrier. The discovery by dentists of the usefulness of eugenol in stopping toothaches likely goes back many centuries. More later about related tooth anatomy and physiology.

Like many herbal medicines, eugenol's history can be traced to antiquity, where it is recorded that Chinese officials chewed on cloves leaves (the source of eugenol) in 266 B.C.E. When trade with Southeast Asia began in 176 C.E, the first spice to be imported into Alexandria, Egypt was eugenol, a testimony to its highly esteemed value. Originally the evergreen cloves tree was only found in Indonesia and the Southern Philippines. The volatile oils of eugenol are extracted by distillation: a separation process, where the buds and leaves are heated until eugenol vaporizes off and condenses into a liquid.

Scholars suspect the ancients were drawn to cloves by it strong aromatic scent, regarding it as a panacea for almost all ills. Today eugenol enjoys wide use outside of dental cements, including flavoring in mouthwashes, as a mild germicide in toothpastes, and in aftershaves and perfumes.

While trying to think of a substitute for the usual zinc oxide-eugenol temporary cement, I had an inspiration. I recalled that some skiers used a zinc oxide paste to cover their nose and lips to block the dangerous high altitude ultraviolet rays. I thought, maybe this paste could be an effective short-term seal for the temporary cap. Conveniently, most ski shops carried the paste among their assorted sun block preparations. With high hopes that the local ski shop had some, we dashed out of the hotel, across the plaza and into the store. The young salesman smiled quizzically at my breathless request for zinc oxide. To our great relief he had just what we wanted.

Back in Dennis's room, I sealed the cap on his tooth with the zinc oxide paste. But I told him to expect the paste to wash out over time, since without eugenol it won't harden like regular temporary cement. To our mutual satisfaction, the cap held for the rest of the day, protecting him from the cold air. Fortunately he was able to make an appointment with a dentist for the following morning in the town of Taos at the base of the mountain. Dennis subsequently told me how impressed the dentist was with the emergency repair of the cap, and simply put it back on with regular temporary cement.

Dennis certainly appreciated my help, which he expressed effusively at the time. But retelling the story always lights him up, especially to a room full of skiers. It's a good human interest yarn that people enjoy. When I am in attendance he likes to ask me to stand up at the end of his talk. Though I admit relishing the attention, I also feel somewhat embarrassed. He tops off his remarks by saying that the best thing I did was save the skiing day for him. This always provokes cheers and applause.

~ * ~

It is now commonly understood why the cold air caused Dennis's tooth to hurt. In the past, this question also puzzled dentists until an extracted tooth was sliced open and studied. Credit for the first micro-scopic examination of internal tooth anatomy goes to Anthony van Leewenhoek, in 1678 Holland. People incorrectly presume he invented the microscope. Though the exact origin of the first microscope is vague, authorities believe it was invented in 1595. Leewenhoek is recognized as the father of high-power microscopy, increasing magnification from the standard of twenty times to two hundred times. His construction of around five hundred scopes is testi-mony to his untiring efforts to improve the instru-ment. Many consider him the greatest of all early microscopists, as well as the founder of microbio-logy.

From these early beginnings, dental scientists were eventually able to unravel the mechanism of tooth sensitivity. It wasn't difficult to exclude the tooth's thin outer coat of enamel for conducting sti-muli. After all, enamel is ninety-six percent inorganic substance, and it was obvious that like finger and toe nails enamel can't feel anything. The early work of embryology determined that skin and enamel are both derived from the same primary germ layer called ectoderm. Thus enamel is actually the outer skin of teeth. Skin's protective covering of the body, with its many variations to accommodate special locations, is certainly a wonder. Intact skin is a barrier to bacterial invasion, and so is intact enamel. The very hard enamel barrier both protects the functioning tooth from physical injury, and penetration of the decay causing bacteria. As a dental student in the early 1950's, I can testify to the enamel's hardness, since at the time we lacked instruments capable of drilling through it. The usual method to cut enamel was with a mallet and chisel. To the modern high-tech attuned ear, this might sound primitive, but it worked. We selected a weak spot on the tooth (commonly an area of decay) and placed the chisel along the crystalline planes of the enamel rods, which extend vertically to the surface, and with gentle blows peeled off the enamel.

The calcified structure under the enamel is entirely different. It is called dentin, and provides the bulk of the tooth. It seems hard like enamel but is actually softer, composed of only seventy percent inorganic substance. No doubt dentin is truly alive, for tooth sensitivity begins here. The pain messages are transmitted by the dentin to the pulp (the living tissue within all teeth, often referred to as the "nerve") through thousands of microscopic tubules that run from the dentin surface under the enamel to the pulp. All the tubules contain living tissue and fluid along with occasional nerve fibers, which promptly respond to most stimuli: thermal changes, chemical irritants, penetration of decay, and trauma.

Now it's clear why Dennis's tooth became sensitive when the cap fell off. His dentist had appropriately stripped off the tooth's enamel to create space for the gold crown, uncovering all the dentin. Without the temporary cap the entire exposed dentin surface was like an open wound. The slightest stimulus caused considerable discomfort. Replacing the temporary cap with zinc oxide paste insulated the dentin and the toothache stopped. Think of the cemented temporary cap as an enamel substitute.

2.

CAVITIES

Everyone knows the term cavities. Just think of the cliché in common usage, "Look mom, no cavities."

But I am sure most are unaware that the technical term for cavities is *caries*—derived from Latin meaning rot or rotten. Furthermore, it seems that many people don't really know what causes cavities, and some even harbor strange myths surrounding the subject. I have repeatedly heard the self-defeating comment that, "I knew I didn't have a chance to have good teeth since everyone in my family lost all their teeth early in life and now have dentures."

Actually, dental decay is a specialized infection caused by bacteria. The prime etiologic (causative) bacterial agent is Streptococcus mutans, and to a lesser degree Lactobaccilli and Actinomyces organ-

isms, all of which are usually among the many different bacteria found in our mouths. They are part of
the normal digestive bacterial ecology. Cavities become a problem when the oral conditions are shifted
to favor the overgrowth of the decay bacteria. Basic
to the metabolism of these bacteria is sucrose. If
there is no sucrose to feed the Streptococcus mutans,
then the bacteria won't thrive, and thus no cavities.

A seminal experimental animal study in 1959, at
the National Institute of Dental Research, conclusive-
ly linked the essential role of bacteria to the initiation
of dental caries. They first showed that a diet rich in
sucrose always led to dental caries in experimental
animals. Next they bred a group of these animals in
gnotobiotic (germ-free) cages, creating a unique co-
lony totally free of oral bacteria. When this germ-free
group was fed the same sucrose-rich diet they all
dramatically failed to develop dental caries. It was an
elegant study, which I had the benefit of viewing
while attending a seminar at the Institute.

One of the more fascinating epidemiological
dental caries stories starts on the isolated Tristan da
Cunha Group of islands in the South Atlantic, lo-

cated fifteen hundred miles from nearest land. The islands were discovered in 1506, but not permanently settled until 1810. There were only fourteen people living there in 1826 when it was absorbed as part of the British empire. Tristan da Cunha soon passed into obscurity until volcanic eruptions in 1961 forced the entire population of two hundred ninety to flee to England. The immigrants reported that their diet on Tristan was varied and nutritious, with vegetables grown in fertile soil, supplemented with poultry, cows, sheep, and abundant seafood. On arrival in England, dental examinations found they were remarkably free of dental decay. We now understand that their healthy, sugar-free diet also prevented decay since without sucrose the cavity forming bacteria couldn't survive. But once the group started indulging in the sugar-rich English diet, their resistance to decay vanished and they developed troubling toothaches for the first time. I don't know whether the toothaches played a role or not, but after a surprisingly short stay in England, they returned to Tristan da Cunha in 1963, when reports reached them that all volcanic action on the island had ceased.

The following are helpful decay preventive measures: conscientious oral hygiene, especially brushing after each meal; considerable reduction in the consumption of candy and all sweets; fluoridation of public water supplies as well as toothpastes; and periodic dental examinations.

Since childhood the importance of brushing one's teeth has been drilled into people. But actually the emphasis could be improved. The outer enamel surfaces of teeth are self-cleansing for the most part, helped by the abrasive action of the tongue and cheek, washed continuously with saliva, and even scraped clean by fibrous foods. It is rare that a cavity begins on the intact enamel, except in those dark narrow spaces between teeth. It is here that dental plaque forms on the tooth, and if not disturbed, leads to cavity formation. So the prevention of decay begins with routine cleaning between tightly aligned teeth, using floss or any of the other myriad devices now available.

The formation of plaque is the first step in cavity formation. Without plaque, there simply aren't cavities. Plaque is a sticky, transparent film, adherent to tooth surfaces, composed of mucin, desquamated epithelial cells, Strep.mutans along with other bacteria, and sucrose. The bacteria actively ferment the sucrose into a strong acid, typically with a pH below 5.0 (pH is the technical measure of relative acidity or alkalinity; neutral is 7.0; further the recordings below 7.0 the stronger the acid; and the higher the recordings above 7.0 the more alkaline). The morbid decay process beneath the plaque is self-sustaining. Once decay penetrates the enamel the infection spreads fairly rapidly through the dentin, where if unchecked the living pulp tissue within the tooth becomes infected, with inevitable serious abscess formation. Early dental treatment will eliminate the decay infection and prevent complications.

A healthy flow of saliva is the body's way of helping with the cleansing. When there is a reduced flow, such as in Sjogren's syndrome, where the lacrimal, or salivary glands, fail to produce adequate secretion, the resultant dry mouth usually leads to an increase in caries. Oral health is at even greater risk following radiation of the mouth for cancer, because total loss of saliva production can occur.

Trying to adhere to a low sugar diet is a worthy effort. But keep in mind that starch, a carbohydrate commonly found in food, undergoes preliminary breakdown into sugar in the mouth by the salivary enzyme amylase. This is another good reason to clean your teeth right after meals.

The urgent need to replace Dennis's temporary cap went beyond relieving his sensitivity. Without the cap the tooth was at risk of developing decay. If, let's say, he decided to put up with the sensitivity until he returned home, he would naturally avoid the tooth as much as possible, including with his toothbrush. Plaque would quickly form on the entire exposed tooth. Decalcification and decay could soon develop on the relatively soft dentin (remember all the enamel had been removed by his dentist). Furthermore, the dentin's thousands of microscopic tubules offer spacious passages for bacterial penetration, accelerating the spread of decay.

Without a doubt, fluoridation of the water supply is at the top of the list of important caries prevention measures. It ranks high among public health successes. On an economic basis, water fluoridation can't be beat: the equipment is inexpensive to install and operate. In addition, the prevention of decay yields an incalculably huge savings in dental expenses, as well as forestalls much human suffering. Contrary to expectations, studies have shown that the fluoride molecules do penetrate the enamel surface and bond to the calcium phosphate molecules—the mineral building units of teeth—increasing the enamel's insolubility by a factor of ten. Swallowed fluoridated water also eventually carries fluoride into the teeth through the blood stream. The era of public water fluoridation began in 1945 in Grand Rapids, Michigan. During the following years researchers monitored the rate of tooth decay among Grand Rapids' almost 30,000 schoolchildren. Eleven years later the decay rate among children born after fluoride was added to the water supply amazingly dropped more than sixty percent. Interesting that it took many years, until 1956, for toothpaste with fluoride to be available. Fluoride toothpastes now offer a decay prevention aid to everyone, regardless whether their water supply was treated with fluoride or not.

Viewed from the current perspective, awash as it is in technological breakthroughs and biomedical engineering marvels, it is hard to fathom why it took many years to discover the role of fluoride in preventing decay. Simply, no one at the time imagined that the obscure fluoride element could have anything to do with dental decay. Scientists only knew fluoride was one of many naturally occurring elements found in the soil. As in many discoveries, the insights and perseverance of one individual made it happen. That man was Dr. Frederick McKay. In 1901, as a recent graduate from dental school, he was troubled about the cause of the "Colorado Brown Stain" on the teeth of many of his patients in Colorado Springs, Colorado. His unrelenting pursuit for an answer led to the eventual correlation of fluoride and prevention of decay. It took many years, but his persistence slowly spread interest among others, including the United States Public Health Service and other key agencies and individuals. With the combined efforts of many, the remote connection between very high levels of fluoride in some drinking water sources and brown tooth stains was eventually established. The pieces now all fell into place. Finally it was understood that the fluoride-stained teeth were also resistant to decay.

The final challenge was to discover whether the fluoride concentration in water could be reduced sufficiently to prevent staining, while still imparting a cavities protective benefit. This link in the lengthy and torturous path of discovery was made. It was determined that fluoride water levels of up to only one part per million did not cause brown stains, while still providing protection from decay. This revolutionary discovery made tooth decay a preventable disease for most people, for the first time in history

~ * ~

A look at history helps to better appreciate the progress made in preventing most dental disease. The literature on dental decay is extensive, and documents this disease's awful impact on mankind from early days. The European Middle Ages—the years roughly between 500-1500 C.E.—were characterized by the remarkable construction of monumental cathedrals. They were astounding engineering accomplishments, a credit to the skill and imagination of the men who built them. But, in contrast, there was universal illiteracy, and the European world was in chaos. A major cause of the turmoil was the ravages of the Bubonic plague. The plague totally disrupted society, killing millions, striking almost every home. The Catholic Church survived as the only stable institution, to which the terrified populace turned for help in droves. Even though folks couldn't read, they understood images and pictures. So with illiteracy prevailing not only among the peasantry, but also within most of the nobility, the Church leaders conceived an imaginative plan to dramatically communicate with their parishioners: they instructed the artisans to adorn the facades of cathedrals with sculptures and base-reliefs. Talented artists were commissioned to create inspiring images depicting the central messages of Catholic faith, along with grotesque gargoyles of animals and figures likely to frighten the superstitious masses. Filling all possible spaces were also masterful portrayals of people from diverse societal levels engaged in recognizable activities. Among these theatrical displays are many stone portraits of a poor soul with a swollen face and painful expression: the look of someone with an acutely infected tooth. Doubtlessly, many paused to focus on this sad image, while helplessly enduring their own dental agonies. Historical records during the Middle Ages indicate that dental infections were common, incapacitating, and occasionally fatal. The sculptures accurately recorded a very troubling rampant dental health problem. I wonder whether the priests of old could have imagined how the survival of these works of art through the centuries would be treasured as tangible models of the past. Like the Elgin marbles on display in the British museum in London, which vividly display ancient Greek life, so in like fashion cathedral sculptures gives us exemplars of those Middle Ages.

The Bubonic plague was arguably one of the worst health calamities of all times. It struck Europe in the fall of 1347 and by the following summer had spread across the continent into England. Estimates of the death toll range widely, but probably totaled at least thirty million during the three hundred peak years of the epidemic. The population was decimated along with major disruptions of the political institutions. Europe was never the same again. The plague arrived when little was understood about disease, and the existence of bacteria was unknown. The popular wisdom, annunciated by the Church, was that God created the plague as punishment for mankind's sins.

Today we know plague is caused by the bacteria Yersinia pestis. How the infection was transmitted is a fascinating and pathetic tale. In the Middle Ages people continually wore the same clothes and seldom bathed. The dirt streets were filled with garbage and excrement, with rats and vermin everywhere. Fleas were ubiquitous: they lived in people's filthy clothing, unwashed hair, and domestic animals were covered with fleas. The thatch roofs in common usage were ideal nesting sites for the rats. Their proximity to humans tragically facilitated the spread of the frightening plague epidemic. Bubonic plague started with the infected rats. The thriving colonies of fleas on the rats would prove critical to the spread of this dread disease. The infection was passed to the fleas in the rat blood the fleas feasted on. When the rats died, the plague-infected fleas left and often attached themselves to nearby humans. Thus the contaminated fleas became the vector for the plague's rapid spread. After a bite by an infected flea, it only took one to six days for the symptoms of the plague to appear. The virulent plague bacteria entered the human circulation where they caused blood vessels to burst. The generalized leaking of blood under the skin produced a typical dark rash. This happened often enough for the plague to be called the "Black Death." Today we can successfully treat the plague with antibiotics. Sadly, outbreaks of the plague still occur in those world areas lacking modern sanitation and hygiene.

The powerful and distressing history of the Middle Ages in Europe can be characterized by the twin scourges of plague and dental infections.

When I started dental school in 1951 the antibiotic age had recently begun. Penicillin was the first antibiotic. The advent of World War II hastened its development to provide a critically needed medication to control infection. Streptomycin was isolated in 1944, and other antibiotics soon followed. Those were exciting times, filled with hope and expectation that these new medications would wipe out dental disease, along with a wide range of other diseases.

Our surgery professor effectively responded to these changing clinical times by illuminating lectures on seriously infected teeth with slide projections of patients with swollen faces—like those on cathedrals. Prior to antibiotics, he explained, those patients required surgical treatment under general anesthesia, and some, nonetheless, experienced serious complications, even life-threatening. He forecasted that primarily because of penicillin we would rarely see such extreme cases. Time proved him correct. But individual dental neglect still leads to serious infections, emphasizing the value of prevention.

3.

SCHOOL

It's important, as well as humbling to keep in mind that each generation is limited in its understanding by what is known at the time. In health care, in particular, we tend to cringe at the wrongheaded treatments and theories of the past. Among the more notable medical errors was the practice of bleeding sick people. The origins of this treatment can be traced to the vague prehistoric Greek world. It's staggering to realize that intentional bleeding was standard medical practice for over three thousand years. Even George Washington's death at age sixty seven in 1799 was probably hastened by repeated bleedings, for what is now believed to have been a bacterial infection. Today bleeding patients is reserved only for unusual disorders, such as polycythemia vera, where the body produces red blood cells in dangerous excess. A meaningful link to the Greek past is the medical term *iatrogenic*—meaning doctor caused complications or undesirable outcomes to treatment—aptly assigning physician responsibility.

Let's consider the (now recognized as archaic) dental practice of what was called "cavity toilet." I don't know where the term originated, only it was routine treatment during my years in dental school in the early 1950's. The need for "cavity toilet" relied on incontrovertible scientific evidence. It was based on the alarming microscopic findings of bacteria in the dentinal tubules. The decision to kill the bacteria was thought essential to prevent further caries formation. In "cavity toilet," after the removal of all obvious decay, before the cavity is filled, ammonical silver nitrate is applied deep within the tooth. The clear solution of silver nitrate is wiped across the cavity floor within the tooth on a small pellet of moistened cotton, and then followed with a second cotton pellet moistened with eugenol. The free silver precipitated by the chemical interaction of the two liquids, impregnates the tooth, turns it black, and kills residual bacteria within the dentinal tubules. This black stain was considered the hallmark of proper treatment. Eventually, when the unforeseen hazards of "cavity toilet" were discovered, this old doctrine was appropriately discredited. Little did anyone know that the silver nitrate could kill the underlying living pulp. My sense of relief with the eventual discovery of this harmful secondary effect was diminished by the recollection of the harm I might have unintentionally caused.

Oral biology gained momentum in the exciting period of the 1950's, and dramatically shifted the traditional emphasis from dentistry's mechanical treatment techniques, to a wider appreciation of the living tissues investing the oral cavity. The tolerance of the living pulp within the tooth to the trauma of routine restorative procedures and to the decay process, in particular, occupied much study. After the halt of "cavity toilet," the lingering concern about the risk of decay by the bacteria left in the dentinal tubules was soon clarified. Evidence mounted that the living dental pulp had more of an ability to defend itself against encroaching decay than suspected. When stimulated by decay, the pulp could deposit a dense calcific barrier, that blocked bacteria and their toxins. Additionally, and most significantly, studies showed if decay was removed before it caused pulp inflammation or infection that the potentially troublesome bacteria remaining in the tubules would die...The "cavity toilet" theory was finally put to rest.

~ * ~

Sometimes unexpected experiences are surprisingly instructive. That's what happened on a Christmas vacation job in the United States Post Office during my senior year in dental school. Along with my friend and classmate Paul, we were hired to help deliver the overabundant seasonal mail. One morning while loading our mailbags, we were interrupted by a question from an obviously inebriated postal employee. He was an old guy who was celebrating the holiday a little early, while his superiors seemed to ignore his intoxication. Needless to say we were appalled by his condition, which only confirmed our then low opinion of the federal civil service. We ignored him as best we could, but his question got our attention when he asked,

"Hey! So, are you guys dentists?"

I looked up and answered,

"Not yet. We're still in dental school."

None of the other workers paid attention to this evolving dialogue. It was as though they heard all this before, and knew what was coming. Our interlocutor then somewhat contemptuously declared in a loud, alcoholic slur,

"I'm going to shlow ya sompt'n ya neverf saw"

Caught in his intense stare and provocative declaration, Paul and I gave him our full attention. Without another word, he proceeded to perform a truly remarkable feat. He took one end of a yellow pencil, placed it between his back teeth on the left side, and broke it in half. That in itself wasn't remarkable, but what came next was. He opened his mouth, revealed full upper and lower dentures, removed the lower with both hands, and held one half of the split denture in each hand. I was astounded. It seemed unbelievable that the broken denture could function at all, let alone be stable enough to break a pencil. Actually, breaking the pencil still would have been impressive even if the lower denture was intact. Most likely a full lower denture would lift out of place if a pencil was held firmly on one side of the mouth. It was astonishing that with his old, ill-fitting, broken lower denture he could perform this trick. I had to accept that despite these limitations, including his drunken state, our postal friend did indeed manage the unlikely. My understanding of the human capacity to accommodate and overcome a difficult challenge was considerably enhanced.

For most people, full lower dentures are notoriously difficult to learn how to control: the narrow mandibular ridge provides little surface area to support the denture, and to make the situation worse, the jaw is curved, forming a horseshoe shape. Stability of the denture is constantly challenged by the powerful active muscles of the tongue and cheeks which tend to displace the prosthesis. The full upper denture, on the other hand, is easier. Its seat against the roof of the mouth provides a firm unmovable base, which contributes to denture stability. What we had witnessed at the Post Office was clearly unusual. Yet I do confess, the old guy had a dramatic flair. The image of him holding the two halves of the lower denture stayed with me, clearly lasting through the years to this very day. It was a bizarre experience, strangely unsettling. There in a dusty back room, as remote as you can imagine from the antiseptic environs of the school dental clinic, we had been given a lesson in denture prosthesis. For the first time I confronted the reality that no matter how well fabricated and fit, the individual patient had to learn how to manipulate their false teeth. They were on their own. Some unfortunately never succeeded, while others did amazingly well. In current years, the reliability of metal implants surgically placed into the lower jaw has improved to where they can create stability for that troublesome lower denture. It seemed that the best approach to those learning to control new full upper and lower dentures was to counsel patience. With time and motivation, people had the potential to master the extremely difficult. That Christmas vacation job in the post office proved more valuable than the few dollars I earned.

~ * ~

The most troubling encounter in our freshman year of dental school was with the human anatomy lab.

Tensions peaked on the first day, as the entire class of one hundred thirty anxious neophyte dentists reported to the huge dissection room, with its high vaulted ceiling. We were not surprised when asked to line up alphabetically, since from the first day of school that was the routine, in classrooms and laboratories. As a result we got to know some people very well, practically living together for four years, while others someplace down the alphabet were nearly strangers. Soon our long line stretched out the door.

I can still remember my first gut reaction of nausea to the awful pungent odor that assaulted us, as we entered the anatomy lab. The smell was overpowering, and permanently dominated the air. With time I somehow got accustomed to it, though always aware of its presence. I soon learned the odor came from the open vats of liquid formaldehyde preservative, used to wet the white cloths we recovered the cadavers with at the end of each session. Not only was the odor sickening, but the solution looked disgusting: It was filled with small specks of fat and other nondescript particles. After a quick look around, and a whiff of the ambient air, my friend Marty fled, and refused to show up for the next two weeks. After weighing the probability of dismissal from school, he finally reluctantly agreed to come to anatomy lab. My offer to dissect for him, so he wouldn't have to touch the cadaver, helped with the decision.

We knew from upper classmen, that the first order of business would be to divide us into groups of six. The peeling off of six at a time from the line went quickly. Each group was marched into a huge walk-in storage refrigerator, where gauze wrapped bodies (like mummies) hung on the wall. Each group took one body down and carried it to an assigned dissection table. Quiet prevailed. Everyone was absorbed with their own thoughts. Standing way back in the line I watched the slow process, hoping we got a slim cadaver, since without Marty my group had only five. As luck would have it, we ended up with the fattest body in the room. He was hugely obese. We just barely made it to our table. Grasping for some levity, we nicknamed him adipose, the technical name for fat.

The standard attire for this lab was a long-sleeved white cotton operating gown tied in the back, with rubber gloves covering our sweaty hands. The small black dissecting kits we each carried added to the morbid intent. This was not going to be another lecture, but a time for us to actually dissect a cadaver. The tension on that first day was almost palpable. We were confronted with a room full of dead bodies, a first encounter with death for many. Moreover, the unstated anxiety about actually dissecting the cadavers seemed to weigh heavily on everyone. Some joked but their laughter sounded strained. Most simply were silent, looked very serious, and acted unnaturally restive.

That first day was memorable in many ways. Our introduction to Professor Miller was also unexpected. He was a fine looking elderly physician with a decided limp, the cause of which we never learned, and was in charge of the lab. Curiously, he was always accompanied by his big, well-behaved German Shepherd dog, who never left his side. Undoubtedly the dog was more relaxed and at home in the lab than we were. To our surprise, Doctor Miller never involved himself with the dissections. But at the beginning of each three hour session, he personally took charge of and led the class in singing old, up-beat songs, accompanied by the musical instruments some of our classmates were encouraged to bring to the lab. At first we were all amused at this bizarre activity, enhancing the pervasively surreal atmosphere. But after a while we recognized it as Doctor Miller's strategy to lighten the mood and help us overcome our squeamishness and reluctance to handle the dead.

I think it did.

4.

NEURONS AND ACUPUNCTURE

When thinking about valuable discoveries, our immense gratitude must go to Alfred Einhorn, who synthesized the first local anesthetic in a chemical laboratory in Germany in 1904. This marvelous drug, which for the first time allowed for painless dental treatment, was called Novocain. In a curious development, the word Novocain has survived as a generic term. It is used by patients and dentists alike to refer to local anesthetics, despite Novocain's replacement with many safer and more effective preparations.

Without the blessing of local anesthesia modern dentistry would be impossible. Howard R. Raper in his book 'Man Against Pain' expressed it well when he wrote, "What life was like without anesthesia, is awful to contemplate."

These days painless dental treatment depends on routine intraoral injections of local anesthetics.

The wonder of these drugs is that they work locally, not remotely in the brain. Most local injections are close to the teeth being worked on, where the anesthetic stops painful nerve transmissions from the teeth. The pain *neuron* (the authoritative name for nerve, and the morphologic and functional unit of the nervous system) propagates an electrical impulse when the level of tissue injury reaches a triggering threshold. In other words, when the dental patient undergoing treatment would start feeling discomfort, were it not for local anesthetics. The afferent (towards the brain) electrical pain impulse travels along sequential *neurons*, bridging the gaps between them (called synapses) with the chemical mediator acetylcholine. For an individual to experience dental pain the neural electrical impulse from the tooth must reach the brain, where it is processed and interpreted. Reliably, local anesthetics block the pain messages right there in the jaw, and efficiently prevent pain impulses from reaching the brain. Thus dental treatment is painless for the awake patient seated comfortably in a lounge-type dental chair. Furthermore, dentists conveniently have choices of local anesthetics with different durations of effectiveness to match treatment needs.

~ * ~

If one considers how focused dentists are in controlling and preventing pain, then acupuncture would not seem an unlikely subject of dental interest. When I became aware of the stir about acupuncture I was enthusiastic. I speculated that it might help control the dental pain I dealt with as a root canal specialist.

The story of my involvement is worth telling, so let's start at the beginning. It goes back to the astonishing newspaper article, published in the Monday, July 26, 1971 issue of the New York Times entitled "Now, Let Me Tell You About My Appendectomy in Peking," by James Reston. Thought of as the Dean of the journalistic profession at the time, Reston had gone to China to interview Chou En-Lai, Premier of the People's Republic of China. He reported that while there he developed an acute appendicitis, and the surgery was performed by a skilled Western-trained physician. When the usual postoperative pain emerged he was surprised to be offered a choice of management techniques: pain medications or acupuncture. Though he knew nothing about acupuncture he opted to try it, unable to resist his ingrained journalistic curiosity. To his immense relief and fascination he was delighted to say the acupuncture promptly stopped his pain, which never returned. The impact of Reston's widely disseminated article was immediate. By the time that President Nixon traveled to China the following May to reestablish diplomatic relations after a hiatus of many years, interest in acupuncture was wide-spread. When the President returned home, having successfully "opened the door to China," he authorized a blue-ribbon panel of experts to go to China to investigate acupuncture practice. The Chinese government welcomed the group, and promised full cooperation. The American panel included representatives from many clinical medical specialties, including academics in neurological sciences. Shortly after the panel's return from China, I heard one of its members report about the experience. He explained they all initially assumed acupuncture acted as a placebo, or had a hypnotic effect. They believed acupuncture's reported ability to produce analgesia (pain killing) couldn't have any scientific basis. Collectively the panel went to China as skeptics, but all returned as believers. Frederick W. L. Kerr, M.D., Associate Professor of Neurology, University of Minnesota, Mayo Clinic, Rochester, Minnesota, as part of the team, summarized their conclusions **that acupuncture anesthesia was an authentic phenomenon, but said they had no idea how it worked.** The report of the panel stimulated a dramatic change. Research in the Western World accelerated in an attempt to unravel the mystery of how acupuncture produced analgesia.

Despite the continuous use of acupuncture in China for over two thousand years, making it probably the longest documented medical procedure in human history, it lacked scientifically-based evidence to support its effect. Over time the Chinese developed a complex, mystical, non-anatomic, theoretical explanation. They conceived of a network of channels or passages of energy distributed in the body, connecting the superficial and interior portions, which regulate the function of the whole body. In addition, the ancients also discovered some disease symptoms were modified and pain lessened by stimulation of certain spots on the body with needles. These best spots for needle insertion were eventually called *acupuncture points.* Albeit the Chinese theoretical basis of channels is doubtful, the Chinese *acupuncture points* have proved useful. The Chinese theories can be likened to a religious article of faith: either you choose to believe or not. For the Western scientific community their theories were inadequate.

The fascinating story of acupuncture has its roots in Chinese oral tradition going back to the Stone Age, with the first authentic written reference dated around 206 B.C.E, during the Han Dynasty. The first Imperial Medical College, founded during the Tang Dynasty, 618-907 C.E, included acupuncture courses as part of the curriculum. However, during the Ching Dynasty, 1644-1911 C.E, an imperial decree banned acupuncture practice. Nonetheless, its use continued among the masses that had great confidence in its benefit. After the Kuomintang party gained control of the country in 1929, President

Chang Kai-shek's modernization program banned Chinese traditional medicine including acupuncture. Once again the ban failed to halt acupuncture's popular use. When the Communist party gained control of the country after a bloody civil war in 1949, a dramatic policy change led to the encouragement of acupuncture. Mao Tsetung, Chairman of the Communist party, in 1950 authorized the convening of the first National Hygiene Health Conference to deal with the country's devastating public health crisis. As part of this herculean effort they revived and reinvigorated their ancient folk medicine. Colleges of Traditional Chinese Medicine were founded based on acupuncture, herbal remedies, and moxibustion--which treat disease by thermal stimulation of specific areas of the skin by ignited moxa-wool or other substances. The government further decreed during the period of the Great Leap Forward in 1958, that henceforth Traditional Medicine doctors would be integrated into hospitals staffed by Western trained physicians. The melding of these two cultures was successful beyond expectations. Acupuncture anesthesia gained medical acceptance, and the innovative appli-

cation of electro-stimulators with acupuncture replaced much of the prolonged tedious hand twirling
of needles.

By 1975 Western research discovered unknown
wonders of human biology. Incredibly, opiate-like
compounds, identified as pentapeptides, were isolated from the brain and the pituitary gland in particular. The peptide was called endorphin, or internal
morphine, due to its structural similarity to morphine. Unbelievable as it was, science had found that
humans produced their own narcotic. Subsequently,
elevated endorphin blood levels were demonstrated
following acupuncture stimulation, as well as after
vigorous physical exercise. Uncovering the long-
hidden secrets of human biology also found specialized cell structure receptors to which opiates must
attach to produce their characteristic effects.

The story of how acupuncture worked proved
more complex than imagined. Many different biological events were identified. The following are the basic responses initiated by acupuncture stimulation: 1.
descending (efferent) nerve transmissions from the
brain block painful nerve impulses coming from

damaged tissues; 2. Significant changes in blood chemistry and complete blood cell counts occur; 3.skin temperature is elevated; 4. and tense muscles relax coupled with increased blood flow to the tissues. These new discoveries were mind-boggling, changing forever our conception of humans.

My membership in the New York Society of Acupuncture for Physicians and Dentists provided an invaluable source of information and advice. In addition to the Society's monthly meetings, I attended their numerous workshops and seminars, where leading researchers and clinicians from around the world reported on developments and theories on pain management. When it came to learning the rapid twirling of acupuncture needles between thumb and index fingers needed to elicit a good acupuncture effect, my endodontic skills proved an advantage. It was similar to the manipulation of dental files used to clean root canals.

After two years of intense study, I decided I was ready to use acupuncture in my practice, but primarily for control of postoperative pain. My decision was based on the two significant limitations of acupuncture. First, it takes about twenty minutes of intense needle manipulation for analgesia to take effect. Secondly, and most critical, only about seventy five to eighty percent of patients will adequately respond. Thus, since acupuncture doesn't work well with everybody it is unpredictable for clinical use. Furthermore, the degree of pain elimination varied, ranging from minor to profound. In my root canal practice, out in the provinces of Pennsylvania, these limitations were a problem. When compared to local anesthetic's ability to invariably and rapidly produce patient comfort, acupuncture couldn't compete.

Now I'll relate the remarkable circumstances of my first acupuncture patient. In a way it was an awesome experience for me, serving to demonstrate the incredible usefulness of acupuncture, while overshadowing all subsequent treatments.

I was still planning the implementation of my postoperative acupuncture procedures, when I received a phone call from a dentist requesting help with a critical root canal problem. He had started root canal treatment on his wife's lower molar a few days ago and she was in constant pain ever since. I agreed to see her immediately. She came alone, which surprised me considering her condition. As instructed, my staff promptly seated her in an operatory. She looked directly at me as I entered the room, and what I saw was shocking. She was unmistakably very weary, her hair was disheveled, she looked pale and haggard without makeup, and her stylish outfit was in disarray. After days of constant pain she was drained physically and emotionally. Rarely had I seen a patient so debilitated and on the brink of collapse. She mumbled, "I've hardly slept in days, and pain pills don't help."

Besides greeting her, I hadn't said anything else.

Next she added,

"The pain is now worse than ever."

My mind was spinning with alternatives to treatment, when I heard myself say,

"I am going to use acupuncture to immediately stop your pain."

She didn't even look up, but simply nodded her head in agreement and resignation.

Silently and quickly as possible, I assembled the acupuncture equipment. I inserted four acupuncture needles: one in the top of each hand at the highest spot of the muscle in the web between the thumb and index finger, and two were inserted at points on the left side of her face, near the painful tooth. The two hand needles were connected to an electro-stimulator, set to pulse at a tolerable frequency and intensity, allowing a steady twitch of the needles. I simultaneously hand twirled the two facial needles. Neither the patient nor I spoke.

After ten minutes of continuous acupuncture stimulation she started to breathe easier and looked up and offered,

"I feel a little better."

I replied,

"Good."

When a total of twenty minutes had elapsed, she declared with amazement,

"All the pain is gone!"

I stopped acupuncture treatment, removed all needles, and sat back on my stool. She managed a smile, looked around briskly, and ran her fingers through her hair, suddenly conscious of her appear-

ance. Her silence spoke volumes of relief and gratitude.

I might have appeared calm, but, in truth, I was
in a state of shock. The wonder of what had just occurred was staggering. I felt as if I had run a race
where the joy of victory was diminished by my exhaustion and gasping struggle to breathe.

We both sat quietly for a while, looking at each
other, somewhat lost in the marvel of what had just
transpired. She was a remarkable woman. Despite
her pain and exhaustion she, nevertheless, had the
strength to not only confront a stranger, but to accept
an even stranger treatment. Her emotional burden
prior to treatment had to have been intensified. For
me, to have affected the total cessation of her pain
remains, unchallenged, as the most dramatic and fulfilling experience of my career.

Finally focusing on my next move I came to an
obvious, yet surprising decision. She desperately
needed rest, and not be subjected to root canal
treatment at this time. Without hesitation I told her,

"Go home and recover. I'll see you in three days
for root canal treatment. In the interval the pain will

not come back. If in the unlikely event it does, call me immediately."

I sat for a while after she left, feeling deeply anxious about whether my bold prediction of 'no pain for three days' would come to pass. I didn't realize until that moment how tense and exhausted I was. Finally I shook my head, stood up and went to my next patient. But the following three days seemed interminable. I jumped at every ring of the phone.

When she returned as scheduled, I immediately knew all went well. She didn't have to spell it out. I almost didn't recognize her. Her broad smile, smart coordinated outfit, and well-groomed look told the story. Without any prompting, she told me how soundly she slept when she got home. Then with a tone of pride in her voice, she added,

"I can happily tell you the pain didn't come back, as you predicted."

I was more than thrilled hearing her report. My sense of relief was as if a great weight had been lifted from my shoulders. I could have kissed her. She probably saw my face light up, and felt a need for an additional gesture. Her right hand was extended and we shook hands, each covering the clasp with our other hand. It was a rare moment.

Her root canal treatment was completed uneventfully and the pain never returned.

With renewed enthusiasm I now embarked on planning to use acupuncture for postoperative pain.

~ * ~

For those of us who practice *endodontics* (the dental specialty of root canal therapy) there is much satisfaction derived from stopping pain. Though there are occasional root canal diseases which create no symptoms and patients experience few after effects from treatment, much of our time is spent treating pain and infection. Root canal therapy preserves a tooth, whose diseased pulp would lead to extraction, in order to eliminate the pulp within. By entering the tooth through a small opening in its crown, canal therapy can remove the sick pulp along with the infection. Then the canal is obliterated with a special filling so nothing can get through the canal. After all, the tooth is neutral, serving as the calcified vessel in which the living pulp resides. After root canal thera-py the functioning tooth remains in place.

~ * ~

The varying degrees of inflammation in the tissues surrounding the tooth, in the jaw, after canal treat-ment usually takes up to three days to resolve, and for discomfit or pain to subside. The intensity of pain during this postoperative period varies, but could be severe. Routinely patients are given pre-scriptions for pain medications, along with antibiotics when indicated. Since the precise intensity of post-operative pain is unpredictable, patients are advised to initially use a mild over-the-counter pain medica-tion such as ibuprofen. If it proved insufficient, then the prescribed narcotic could be added.

Patients self-selected for the postoperative acu-puncture treatment. They were told to choose how to manage their pain after the local anesthetic wore off: they could use the pain prescription, or return to the office for acupuncture. The choice was theirs. Of the forty five patients who returned for acupuncture, over ninety percent experienced relief from the acupuncture, and the pain did not reoccur. The sample could not be considered as statistically significant, but nonetheless, the patients would all agree it was worth the bother.

I must admit I was an early skeptic about the claims for acupuncture. It wasn't until I treated my first pain patient that I became convinced of acupuncture's merits.

With the passage of time, acupuncture has established a viable, yet minimal clinical niche in the American health care system. There are a wide range of human ailments for which acupuncture can offer meaningful relief. But like all procedures, acupuncture will succeed only with defined targeted problems, and best not viewed as a panacea.

5.

PATIENT EMPOWERMENT

Pain is a terrible experience. People often are able to deal calmly with complex health problems, but pain can unsettle and frighten. Pain is a personal subjective experience, almost impossible to recognize unless the patient shares the knowledge. Some of my patients eagerly tell me they are hurting, and others try to evoke sympathy, or dramatize their distress (most often women) by crying. The non-communicative stoics are just the opposite. After some urging, their calm announcement that they are in agony always startles me.

How people deal with pain might widely vary, but underneath they are psychologically and physio-logically similar. Building on these similarities I for-mulated a pain control method called *patient empo-werment* that proved to be very helpful. In general, when patients became better informed about treatment and learned what to expect during the postoperative period, they experienced a reduction in anxiety which in turn lessened postoperative pain. The approach took maximum advantage of patient's inherent potential to marshal emotional and physical resources to reduce symptoms and improve healing. The benefit of focusing patients in a positive and receptive way towards healing was expressed well in the following quote (regrettably the source is not available):

"...the way we think about things determines how we feel. Our feelings are then manifest in the limbic area of the brain, which stimulates the neuroendocrine-immunological link."

I am talking about the astonishing link between mind and body. For the longest time it was believed that the body functioned independently of the mind. We now know the brain's neurologic signals cause hormonal and chemical secretions, which in turn alert immunologic mechanisms. These immune reactions are the body's essential defense against infection and disease. AIDS is devastating because it prevents the immune system from working. Moreover, our brains continuously mold during life, shaping circuits which connect our emotions to the release of useful hormones and chemicals.

The other side of the mind-body connection is the impact of disease on the brain, causing the brain to generate negative thoughts. Therefore, positive, hopeful thoughts improve health by balancing the negative imputes of illness. *Patient empowerment* succeeds because it promotes positive thoughts. Naturally, positive thoughts alone can't cure disease, but they do assist in recovery from illness. For example, it's not uncommon for patients not to recover as expected following surgical operations if they slip into depression. In contrast the optimistic patient, with the same operation, is out of bed as soon as possible, strolls the hospital corridors, and is impatient to go home. Health and disease are influenced by a complex of variables, some environmental and some individual.

Words are the instruments that reach the brain and are the power brokers of thoughts and thus life. Selecting the right words and using them in a positive constructive way is a tremendous challenge. I admit to have struggled with this skill. Only through trial and error did my plan of how to help patients deal with anxieties and postoperative pain take form.

In the *empowerment approach* patient's fear of treatment was assumed. Telling people not to be afraid of root canal treatment proved a waste of time. Most patients were fearful of the unknown. The opportunity to relieve this fear begins before treatment, and is the first stage of *empowerment.* While sitting on a lowered stool—an incredible reduction in patient tension is achieved when patients look down at the seated dentist—a preliminary dialogue (not a lecture) is held. This first encounter is important. The last thing patients need were to look up at the authority figure of the dentist while they sat in the dental chair, rigid with tension. When patients look down on the dentist, they become empowered. The amount of time spent in discussion varies, adjusted to meet the needs of each patient. For most people, fear is lessened by clarifying what was wrong with their tooth (simple drawings often assist), how treatment would be comfortably rendered, and how pain—high on the scale of anxieties—would be controlled. A projection of optimism and encouragement is basic.

The second stage of *empowerment* is the most tangibly productive. By that I mean, though the improvement in how patients deal with pain is difficult to measure, their attitude to the pain is not. Let's put these thoughts in perspective.

In my years of practice before *empowerment*, there was a significant difference in how I dealt with postoperative pain. When treatment was finished, I promptly left the patient in my haste to go to the next patient. In parting I said that prescriptions and instructions would be given by the dental assistant. I usually enjoyed an emotional lift after finishing a root canal treatment, a feeling of a job well done, and a responsibility fulfilled. The pills, I assumed without much thought, would deal with the patient's postoperative experience. For some it was enough, but for many it was totally inadequate. The reality was that with little understanding of what to expect, many patients were very apprehensive.

I have often thought the second stage of *empowerment* the most important. To maximize *empowerment* a multifaceted approach works best, which means building on every possible asset. I decided it was important to stop delegating the patient's final care to the assistant. I am convinced that my personal involvement favorably influenced the patient's recovery.

If I can select the most useful information in reducing patient anxiety, it was to place a time limit on acute postoperative symptoms. Patients were universally relieved to learn that within three days the acute symptoms would fade. I explained that three days were needed for the body to recover following removal of the sick pulp, and the elimination of the infection. The reduction in patient anxiety along with an elevation in their comfort level, was gratifyingly obvious.

It is important to understand that the three-day period for recovery wasn't simply an arbitrary number. When the morbid root canal disease is successfully treated, the biological response changed from sickness to repair, usually within three days. If by chance a complication occurred and healing was delayed, then the informed patient was better prepared to recognize the problem and inform the dentist. By sharing the critical parameter of time with the patients, they became knowledgeable partners, whose responses were valuable in judging therapeutic outcomes.

Moreover, patients learned that total healing of the internal tissues around the tooth could take many months. Therefore, some tenderness (not pain) in the area is to be expected during this prolonged period.

Just before I said goodbye, patients were requested to call the office the next day to report progress. I always explained that I was a worrier and needed the assurance they were starting to feel better. My expression of concern was a simple gesture, but by insisting on a follow-up my continuing involvement in their recovery was emphasized. They were not left alone.

I instructed my receptionist to keep a record of the calls. If the patients were doing well, there was no need to call again. On the other hand, if recovery seemed slow, then they were requested to call again the next day. Of the five hundred forty patients who called—during an early trial period of *patient empowerment*—ninety percent reported improvement after twenty four hours, while the remainder took a day or two longer. These well-informed patients characteristically sounded optimistic, and said the pain was subsiding on schedule. In the past, before empowerment, some patients called on their own initiative the day after treatment, anxious and worried why they still had pain after twenty four hours. There was a dramatic difference in postoperative responses between the pre and post *empowerment* patients.

6.

MICROSCOPES AND HAND MIRRORS

The unexpected happened one day while reading a dental publication. I spotted an announcement for a program to be held at Harvard Dental School, in Boston, on September 25, 1982. I could hardly believe what I read. A dental operating microscope, called the Dentiscope, was to be introduced. After many years of endodontic practice, I could only imagine what it would be like to work through a microscope. I was excited and immediately phoned the school to sign up.

For a long time, I knew neurosurgeons, and others had been operating with the aid of surgical microscopes, and I was envious. After all, when I compared where and how I had to operate—within minuscule root canals—I figured I was a micro-neurosurgeon, and needed a scope. Using a microscope to enhance vision during surgery was not new. Back in 1953 the Carl Zeis Company—then of West Germany—had developed the first commercial binocular operating microscope, and its use had spread rapidly through all the surgical disciplines, except dentistry. It appeared that dentistry's time had finally arrived.

The two co-developers of the Dentiscope, Apotheker, D.M.D, and Jako, M.D, along with the representative from the Chayes-Virginia Dental Supply Company, the manufacturer of the scope, were at Harvard to greet the (disappointingly) small group of five dentists who had enrolled. Nonetheless, their lectures were fascinating and instructive. The exciting part of the seminar for me was the opportunity to work through a scope on extracted teeth. I knew immediately that the scope's visual enhancement, achieved with a fixed amplification of seven times and supplemented with adjustable built-in fiber-optic lighting, would be of significant clinical value in endodontics. The focal distance of nine inches positioned the scope above patients with sufficient clearance not to intrude into their personal space. My order for a scope delighted the rep from Chayes-Virginia. As if to be sure I didn't change my mind, he arranged for its prompt delivery in two weeks.

The full incorporation of the scope into my nonsurgical and surgical root canal practice took time, and as expected (and hoped) led to treatment improvements. Though the scope came late in my career, the welcome changes were professionally stimulating. The arrival of the microscope marked a refreshing turning point, creating new challenges and unanticipated rewards.

I never understood why it took many years for the operating microscope to achieve acceptance by endodontists. But ultimately it use did spread, leading to a reshaping of clinical practice. The wheels of change started to move rapidly in 1996 when the Commission on Dental Accreditation of the American Dental Association ruled that microscopy training must be included in Advanced Specialty Education Programs in Endodontics.

The dramatic change from casual interest in microscopes to fevered involvement heralded a new era in endodontics. Microscopy in endodontics came of age, but its gestation period was unduly long.

~ * ~

Now that I have dealt with the high-tech side of my profession, hopefully creating the impression we are up-to-date, I will turn to the purely mundane and talk about the dental mirror. You will be surprised to hear that it is complicated to use and irreplaceable. Patients don't pay any attention to the dentist's mirror, but for dentists it is an absolutely indispensable instrument. Believe me; the mirror is actually more important than the scope. The scope allows us to see better and thus perform more accurately, but its use depends on the mirror. Why this is true will become clear as we go along.

It is obvious that direct vision into the mouth is limited: Only the front teeth and a partial view of others can be seen. The mirror overcomes this limitation by providing a clear indirect look into all those dark corners of the oral cavity. By the way, learning to use the mirror was by far the hardest dental skill I had to master. Looking into a mirror to guide your work is hard enough, but the difficulty is compounded since the image in the mirror is backwards. That is to say, in addition to adjusting to working indirectly through the mirror, I had to remember to move my instruments in the opposite direction to where the mirror showed. Eventually, with persistence the technique was mastered. Clearly motivation is the answer: It generates determination, which in turn makes the difficult possible. Since I wanted to become a dentist, there wasn't any choice. Eventually, like generations of dentists before me, I was able to work comfortably while keeping my eyes on the mirror.

Before I go any further, I don't want to overlook the other indispensable uses of the mirror: it directs light where needed; and reflects tissues out of harm's way. The mirror shares the burden of stretching the cheek for access into the mouth, and can safely push the tongue out of the way. Don't forget it's quite dark in the oral cavity, so very bright illumination is needed. The overhead dental lamp is the powerful souce. Its suspension on a universal hinge allows unlimited focusing movements. Here's where the hand mirror comes in: The beam of outside light is aimed at the mirror in the mouth, which in turn redirects the light. Sounds tricky, but is doable.

When I decided to become a dentist I knew manual dexterity was essential, but to what extent came as a surprise. I never realized the proper use of the dental mirror could be so challenging. The complexity flows from the synchronization of the mirror's three functions: vision, illumination, and reflection of tissues.

The mirror, held in a three-fingered pencil grip, is inserted into the mouth, while the other two fingers reflect the cheek and find a secure rest on adjacent teeth for stabilization. Next the focus of the mirror is adjusted to get a clear image of the tooth to be treated by the three fingers holding the mirror, and then firmly stabilized in this position. We aren't finished with the mirror. Next is illumination. Usually the dentist reaches up to the overhead light with the other hand, to aim its beam onto the steadied mirror. As you might imagine when patients move or the dentist wants to see a new spot, the mirror and light are changed. All these maneuvers of hand, focus, and illumination occur repeatedly during treatment, and are executed automatically, without difficulty by an experienced dentist. Hope you're impressed.

~ * ~

When the microscope was introduced, nothing changed, and everything changed. I still looked down from above the patient onto the mirror, but now through the scope. Changes were needed with the scope. The first modification was unexpected. The hand mirror's reversed image is itself reversed by the scope. A surprising effect, which I can't explain other than to say it's due to the scope's complex optics. Unlike in my student days when I had difficulty learning to use the mirror, this new adjustment came easily. Before long I was working smoothly, even switching between looking through the scope and working directly without it, as of old. I was reassured that an old dog (read dentist) can indeed learn new tricks.

A welcome feature of the microscope is the built-in fiber-optic light, which is projected through the scope. The light, therefore, automatically targets the mirror as the scope is focused. The overhead dental light is no longer needed, is turned off and pushed aside. Life is made easier.

The final adjustment took some thought. It concerned the passing of instruments. For years I had an efficient system of working with an assistant, called four-handed dentistry. I sat on one side of the patient and the dental assistant on the other; instruments were passed to-and-fro quickly and smoothly. Above all else, success depended on my peripheral vision. Even though I kept my head down, focused on the mirror in the patient's mouth, I could see widely without turning my head. The instrument exchanges were made out in front of the patient, within easy reach. Thus, treatment was completed within the shortest possible time, with minimal conversation with the assistant. Four hands worked as one, with my eyes continuously on the patient. Naturally, the patients were grateful for the speedy work.

Unexpectedly, the scope disrupted our instrument passing ritual. Since the binocular operating microscope used both eyes, my peripheral field of vision was reduced to where I could not see the passed instruments as usual. All maneuvers on my part to see the passed instruments became clumsy and time consuming. With my peripheral vision narrowed I had to turn my head away from the scope to locate the passed instrument. These initial trials were unsatisfactory. Before long we arrived at an effective solution. The dental assistant had to deliver the instruments differently: they had to be within my new narrowed field of vision. To accomplish this change she had to sit closer, reach much further, and learn to judge where my new field of peripheral vision extended. For my part, I could no longer move my hand out in front of the patient, somewhat towards the assistant, when ready for an instrument exchange, but had to place my hand close to the scope within the confined field of vision. It might sound simple, but this change was awkward for both of us. At first we found ourselves reverting to old moves before the new technique was mastered and could be executed perfunctorily as usual. The adjustment was worth the effort, and before long we were back to a rapid, seamless four-handed routine.

Micro-endodontics eliminated much of the invisible, which led to improvement in many aspects of therapy. For instance, during root canal treatment before the scope, there was difficulty illuminating the inside of teeth through small access openings. On the one hand, efforts to conserve tooth structure called for making small openings. But on the other hand, the need for better vision often required larger openings, with unavoidable structural weakening of the tooth. The scope's magnified and brightly illuminated image made possible successful treatment with smaller openings. Also, the improved vision inside the tooth disclosed the internal anatomy with more clarity, leading to easier location of the (very minute) root canal openings. The scope also proved useful in root canal surgery. Situations occasionally require the surgical elevation of the gum tissue for access to the root tip within the jaw. In addition to removal of abnormal tissue from this boney area, the scope's brilliant enlarged view of the root apex helped to locate and seal minute openings at this end of the tooth. Yes, micro-endodontics not only helped perform more precise root canal treatment, but was also agreeably energizing.

7.

COMPLICATIONS

Unusual complications can be very instructive. They catch you off guard, take the wind out of complacency, inculcate humility, and promote productive concentration. So, even with ongoing improvements in the reliability of root canal treatment and advancements in dentistry, serious disseminating dental infections do occur.

Take for example the following. In the 1960's I treated a serious case of root canal infection which had spread into the maxillary sinus. The patient presented with facial swelling and low grade pain in the upper jaw. A thorough examination, including periapical radiographs of the area, found root canal infections in the adjacent upper left second premolar and first molar. The teeth were firm, and despite considerable gum recession around the molar, the marginal tissues showed no signs of inflammation.

The treatment of the canals progressed routinely, but the facial swelling proved unresponsive. Antibiotic therapy was also ineffective. With the canal treatments finished, it was clear that the fluid accumulation in the tissues needed release. A small incision into the height of the swelling in the tissues above the molar immediately produced a generous flow of pus. To ensure the release of all the pus a drain was placed in the opening to maintain its patency for a few days. I felt encouraged at this point that the prospects for recovery had improved. Unfortunately, after the drain was removed the flow of pus continued unabated.

The combination of treatments at that point had failed. The impasse challenged my self-confidence. I couldn't help wondering what I had overlooked. Unsettled, with niggling feelings of guilt and uncertain about the prognosis, I advised the patient that we should open the area surgically for examination and thorough curettage. Furthermore, due to the complex anatomical area I was somewhat anxious about the surgery. The otherwise healthy and composed fifty-five year old female patient didn't hesitate to agree to the procedure.

What I found surgically was unanticipated. The extent of the damage to the surrounding jaw bone and adjacent maxillary sinus (the large air filled cavity in the cheek bone from which the flow of mucous into the nose helps to maintain its important moisture coating) was startling, way beyond expectation. My appreciation of the destructive potential of infected teeth was meaningfully upgraded.

I found that much of the maxillary (cheek) bone above the teeth had been destroyed, and for a considerable distance on all sides the remaining bone was engorged with blood and soft in texture—undergoing decalcification and resorption. The normal mucosal lining of the inside of the maxillary sinus was swollen, with pus accumulations evident throughout the area. Cautiously, much of the abnormal sinus mucosa was removed (I was aware that with elimination of the infection, the mucosa would reform). The periodontal (gum) pockets around the molar were thoroughly cleaned of all abnormal tissue and packed with some of the soft jaw bone from the sinus area. Hopefully this graft would aid in the closure and healing of these serious gum pockets. The gingival flap was lowered and sutured firmly in position. For protection of the huge wound, the entire area was covered with a surgical cement dressing which supported and stabilized the flap

The patient was monitored carefully for many months, and gratifyingly healed uneventfully. Even with the complex of endodontic, periodontic and sinus diseases, successful repair did occur. Of particular concern were the pockets around the molar, which closed and could not be probed beyond normal depth. I was delighted and relieved with the case's favorable outcome. Frankly, I never would have expected such a good result.

Even though we knew, back then, that infections do spread from teeth into the adjacent sinus, there was a tendency to de-emphasize the potential complications. My confidence in the ability of root canal treatment alone to not only clear infections from within teeth, but also from the surrounding tissues, was probably overblown. I came to recognize that surgical intervention was at times required when healing did not occur after root canal therapy. The question of why surgery helps has been puzzling. Besides the removal of abnormal tissues during surgery, I suspect the surgical trauma also activates an otherwise sluggish latent biological repair process.

Therefore, logical sequential treatment of root canal infections relies on first eliminating the infection within the teeth. By stopping the spread of infection from the teeth into the surrounding tissues, the body's defense system often clears up the residual peripheral infections. If that fails to happen, then surgical treatment should be considered, while still preserving the teeth.

To understand the mind set at the time, a look back is helpful. In the early 1960's Endodontics was in its early days of maturation; to be shortly approved by the American Dental Association as the newest official dental specialty. At that time, possibly too much energy was devoted to convincing the profession that root canal treatment could reliably save teeth. Also most seminars, conferences and discussions focused on techniques of cleaning and sterilizing root canals, along with resolving conflicts over how best to fill the treated canals. We were somewhat myopic about broader issues.

~ * ~

Sometimes the presentation of signs and symptoms of dental infection can mimic or even mask other pathological processes. The next case is a good example of where I was fooled at the beginning. All the usual diagnostic tests were compatible with a root canal infection, including the bone loss around the apex as seen on a radiograph. Convinced of an unambiguous diagnosis of pulpal disease, I performed root canal treatment on the involved tooth.

On routine reexamination one month later, I was shocked to find serious deterioration on a new radiograph: the bone loss around the tooth had expanded; and most troubling, one surface of the tooth's root structure had extensive resorption (dissolution). I concluded the jaw bone infection, which had spread from the root canal, had gained a head start, and was aggressively worsening. I proceeded with a surgical curettage, a routine procedure.

My assumption ultimately proved wrong. On hindsight, I should have recognized that the excessively rapid destruction of bone and tooth were very atypical, and warranted reevaluation before proceeding with surgery.

Surgical exposure of the lesion found a mass of abnormal tissue within the jaw that was unfamiliar and different: The tissue was friable, almost gelatin-like, and hemorrhaged at an alarming rate during its removal. The inability to control the excessive bleeding led to the decision to close the wound before all the pathologic tissue was removed—the residue of which would, unfortunately, cause future problems. A broad-spectrum antibiotic was prescribed and coverage continued for one week postoperatively. All the abnormal tissue removed during the surgery was sent for histopathological examination and diagnosis. The pathologist identified the lesion as a central giant cell granuloma (also known as a giant cell reparative granuloma), classified as benign, and idiopathic (cause unknown). But the benign designation was little comfort, since the lesion is characterized as locally aggressive and destructive, with a tendency to recur after surgical removal. Though it wasn't malignant, it was fast growing and damaging. Since there wasn't any doubt that some of the lesion was left behind, a recurrence seemed likely.

Prior to receiving the pathology report, I wishfully had speculated if this unusual lesion was only infection related, then the fragments left behind might be removed by the body's defense cells, and the area would heal. The eventual diagnosis of a rare giant cell granuloma downgraded the prognosis.

Due to the uncertain outcome, the patient was seen monthly for clinical and radiographic examination. Seven months after the first surgery he presented with a recurrence. Doggedly I proceeded with a second more radical curettage, determined to eradicate the lesion while preserving the teeth. At the time, I felt the effort was justified, since the three teeth now involved in the lesion were prominently in the front of his mouth. On reflection, I should have known better. The invasive nature of this lesion led to its spread into inaccessible areas enveloping the teeth, dooming the attempt to remove all the remnants. My stubbornness led to failure of the second surgery.

The surprising amount of bone and root resorption initially detected one month after root canal treatment, should have signaled an atypical finding; and called for additional consultation before proceeding with surgery. Regardless who performed the first surgery, it was necessary to discover the lesion's true nature—the radiograph could not in itself make the diagnosis, it only raised questions. In reality, the first surgery was a biopsy, which took only part of the lesion so microscopic examination could identify it.

I now see that the patient would have been best served by early referral to a maxillofacial surgeon. As it turned out, he eventually was, after the second curettage failed. The subsequent surgery was successful, but three teeth had to be removed to insure the elimination of the entire lesion.

Of course everyone, doctors as well as patients, would prefer a more perfect world, in which a simple test would screen all cases for detection of those threatening invisible lesions. Until such a day arrives, we do our best with the evidence available. The reality of dental and medical treatment is that we often must treat the obvious, hope for the best, and be prepared to deal with the unexpected. Nonetheless, in this case, surgery was a hasty decision.

~ * ~

The masking or mimicking of other unusual jaw lesions by more commonly found dental infections is fortunately rare. It is therefore not surprising that a misdiagnosis could be made by not suspecting a malignancy, especially in the early stages of these uncommon lesions. Malignant neoplasm's of the oral cavity account for no more than three to five percent of all malignancies. Furthermore, it is believed that only one percent of such metastases (spreads) are from lesions below the clavicle.

The following is an example of early diagnostic difficulty, and cites the complications and treatment that ensued with a malignant metastatic carcinoma.

This case was unusual in many ways. It occurred in a local hospital dental clinic, where I was the Director of the Dental Department—a second career after I retired from private practice. The patient was hospitalized for unresolved left lobe pneumonia. During the early days of the patient's hospitalization, he was referred to the dental clinic for assessment of an acute flare-up of pain in his lower left first molar. The dental resident on duty, found the tooth to be both sensitive to percussion and tender on palpation of the adjacent tissues. Advanced periodontal deterioration with inflamed soft tissue extruding from the sulcus on the cheek side of the tooth was also noted. A dental radiograph obtained at the time showed a large area of radiolucency (bone loss) associated with the roots of the tooth. These findings were compatible with the presumptive diagnosis of an acute abscess. Because the patient was anxious to have the tooth removed, it was extracted and the socket thoroughly curetted, without complication. No antimicrobial treatment was prescribed, because the patient was already receiving antibiotics for his pneumonia.

Because the patient's pneumonia continued to be unresponsive to antibiotics, and the radiological appearance of the pneumonia changed, he underwent bronchoscopy. Biopsy specimens obtained at this procedure revealed a poorly differentiated non-small cell malignant carcinoma. The worsening of his condition led to additional tests which found a mass in the head of the pancreas.

About one week after tooth removal the patient complained of sudden increased pain and swelling in the jaw at the recent extraction site. He was promptly referred back to the dental clinic. Along with his considerable pain, he had difficulty opening his mouth. He also said his lower lip on the side where the tooth was removed felt numb. This symptom was of concern, since it indicated pressure on the main nerve serving the jaw. Oral examination revealed a remarkable growth which extended from and was attached to the socket where the abscessed tooth had been removed. The growth looked strangely differ- ent. It projected up from the jaw for an inch or more and prevented the patient from fully closing his mouth. The color and texture of the growth was un- usual. Instead of appearing bright red like new heal- ing tissue with its abundant circulation, the growth looked blanched. A new radiograph showed consi- derable expansion of the destruction of the mandible (lower jaw). Even before biopsy, the clinical findings suggested oral cancer.

The immediate objective at this point was to debulk the lesion so the patient could close his mouth, and submit all tissue for microscopic evaluation. Under local anesthesia, curettage deep into the socket was performed, with removal of a large quantity of fragile, brittle tissue. Secure wound closure was achieved with multiple interrupted chromic gut sutures, which would slowly dissolve and not require removal. The patient was returned to his room and seen daily by the resident for oral examinations. No postoperative complications were encountered.

The pathologist confirmed the presence of metastatic (having spread to the jaw from elsewhere in the body) poorly differentiated carcinoma in the mandible. In an effort to control the cancerous tumor growth, palliative localized irradiation of the mandible was started. During his 4-week hospitalization, tests finally clarified that the patient's primary tumor was pancreatic, and that the lung, stomach and mandibular lesions were metastatic. Sadly, but as anticipated, in ten days following his last dental clinic visit the patient expired.

This case illustrated. an uncommon pitfall in the evaluation of relatively common endodontic disease. It accents the importance of a complete work-up, with attention to clinical history, pulpal health, presence of paresthesia (lip numbness), and growth rate of the lesion. In addition it was confusing both medically and dentally. The case demonstrated once again how important it is to monitor patients closely following treatment. When they don't respond as expected or show unusual signs and symptoms, reevaluation of the original diagnosis is required.

8.

DEAN TIMMONS AND OTHERS

Going to dental school was formative. By that is meant I entered as one person and left four years later as someone else. Becoming a dentist was no mean feat. Those were trying years, filled with anxiety and stress. Yet they were magnificent times. I bonded with others as we shared the endless challenges: we rejoiced in each other's triumphs as if our own; talked endlessly of our futures; and laughed at secret jokes. When we proudly emerged as dentists it was with gratitude and relief.

~ * ~

When I entered Temple Dental School in 1951 we were fortunate to have just the right man as dean. Dr. Gerald D. Timmons, in my opinion, was a truly effective dean, whom we respected and admired. Dean

Timmons was clearly a good administrator and, possibly, a better politician, very approachable, earthy, not at all the staid academic. He was appointed Dean in 1942, coming from the Indiana University School of Dentistry, where he was acting Dean. His first charge at Temple was to plan for a new dental school building to replace the ancient structure on eighteenth and Buttonwood Streets, used since 1896. His role in the development of the new dental school was substantial. The actual purchase of the building from the Federal Government's War Assets Administration, at a considerable discount for only three hundred thousand dollars, was delayed until the end of the war years. According to histories, Dean Timmons supervised all aspects of the building's renovation in 1947. He visited many dental schools around the country to gain information on design. Furthermore, his leadership must be credited with creating a modern facility, along with assembling a prestigious faculty. Dean Timmons' tenure lasted until 1964. In further recognition of his abilities, he served as President of the American Dental Association between 1962-1963.

Thinking about Dean Timmons, brings to mind my first encounter with him. I was in a musty, not too clean, freshman laboratory, where I had just cast my first technique gold restoration. To restore its usual gold luster, the blackened casting was cleaned by quenching the hot metal in pickling acid. While focusing my attention on the bubbling action of the acid, I failed to notice that Dean Timmons had arrived in the room and was peering down over my shoulder at this process. My first knowledge of his presence was his comment about the yellow acid and the many dark fragments floating in it.

"It looks like tiger piss, doesn't it?" he said.

When I turned, there he was with a cherubic smile. Caught by surprise, and intimidated by the encounter, all I could muster was a feeble grin. He tapped me on the shoulder and left. Even though I had no idea what tiger piss looked like, I speculated that the dark yellow acid might fit the description.

The Dean's unfailing ability to remember first names always delighted people, and his warm, friendly, informal manner, was a natural part of his midwestern personality. A good example was his visit to the Dental School a few years after retirement, and six years after my graduation. I was working as an instructor in the postgraduate endodontic clinic, when I was surprised to hear his familiar voice call out from the doorway,

"Howie, how the hell are ya?"

His broad smile and amiable greeting still had incredible positive resonance. It wasn't as if I was a close friend of his, and I never had more than a few personal encounters, so I was in awe and pleased he remembered my name. I waved, called out that I was well, and said how nice to see him again. He waved back and left. I never saw him again.

But he was a tough taskmaster when he thought our behavior or dress and appearance fell short of proper professional standards. We learned quickly what he meant from his repeated utterance,

"If you want to be treated like a Doctor, then you must act and dress like one. Barbers might go around without a close shave, or wear their shop gowns out to lunch, not Doctors."

Every morning as we arrived at school, Dean Timmons greeted us in the lobby. It became evident he was on patrol, checking to make sure we were properly dressed with shirt, tie and jacket, and in particular, were clean-shaven. Those students with very dark whiskers had to make extra efforts to pass his inspection. It wasn't unusual for someone to be sent home to shave again.

As students, none of us were initially aware that dental education was undergoing pressure to change from its traditional pragmatic mechanical roots, into a biologically based oral health care profession. Dean Timmons was the right man to guide these changes and balance the opposing ideas. As our education progressed, we came to view dental education as having a split personality. Certain faculty members tenaciously hung onto the old ways and were suspicious of the new. The conflict was most evident in the long hours we spent learning how to make full dentures, beginning in the freshman year. The unstated message was clear: Trying to save teeth was futile. While at the same time we learned how root canal therapy could preserve teeth, and periodontal disease wasn't hopeless but could be treated. In time, innovations and scientific breakthroughs did come like a rising flood, to irrevocably change dental education.

Finally, in summary, Dean Timmons' influence in molding a new generation of post World War II dentists was positive, and in important ways offered an appropriately tempered leadership suitable to the times.

~ * ~

The Dental School was strategically located a few blocks from the Medical School and Hospital. Their proximity facilitated the sharing of faculty, to the greater enrichment of the Dental School. Most beneficially was the appointment of John A. Kolmer, M.D., as our professor of medicine. He was a senior, nationally preeminent physician, whose passion and medical erudition held our attention during lectures, reinforcing our understanding and increasing our retention of the material. Dr. Kolmer's lectures during the week were held in the dental school, but his presentations on Saturday afternoon in the hospital amphitheater were the highlights of the week. The gallery was always filled to capacity with students and guests alike. Though scheduled as educational, Dr. Kolmer's delivery was pure theater, which nobody wanted to miss. The classical presentation each week was the same: a live patient who had the disease to be discussed was brought from the hospital onto the amphitheater's podium. The patients were sometimes in a wheelchair, while often they were stretched out on a gurney. In time-honored pedagogical fashion, he taught us how to make a diagnosis by example. His script was basic: he pretended ignorance about the patient's illness, as he would with a new patient. Some patients were nervous and ill-at-ease when Dr. Kolmer asked, in his booming authoritative voice, what was wrong. Others rose to the occasion, and would take over if allowed. A thorough review of the past medical history sometimes proved critical, revealing clues to the ultimate diagnosis or eliminating some possibilities. Next came the physical examination, where he vigorously thumped and probed, much to our delight. He read aloud the findings from the blood tests, interpreted radiological studies, and filled his commentary with illuminating editorial comments. His pursuit of the diagnosis was like an exciting criminal investigation, systematically plodding along, then suddenly arriving at the only possible explanation of the patient's condition. We listened and watched on the edge of our seats with transfixed attention. None of us wanted to miss anything in the enfolding drama. Though short of stature, Dr. Kolmer seemed a giant. He had unquestionable authority, and his strong resonant voice carried the dramatic oratorical delivery throughout the room without amplification.

When I envision those Saturdays in the amphitheater, it is reminiscent of a painting of twelfth century Bologna, Italy, Europe's first Medical School: similar amphitheater; students hanging forward in their seats; and the celebrated professor holding forth. Some good things don't change.

As one might expect, we were proud being Dr. Kolmer's students, and justifiably felt our medical education was exceptional. This opinion played out one day at Fort Riley, Kansas, where I was serving in the United States Army Dental Corps. There were about twenty other recent dental school graduates from across the country, with whom we all engaged

in endless dialogues and debates about whose training was the best. So, I wasn't surprised to overhear
my classmate Stan's heated discussion with Mike,
from Chicago's Loyola Dental School. I paid little
attention, until Stan called out he would prove his
point by asking me a question. I suspected he would
dredge up some obscure fact, confident of my mem-
ory, since we had studied together. Sure enough, as I
turned to face them I recognized Stan's familiar flush
of excitement. Looking straight at me with as serious
a look as possible, he asked,

"Howard, what is the blood test for infectious
mononucleosis?"

I slowly turned while nonchalantly throwing the
answer over my shoulder,

"The heterophilic antibody test, of course."

Stan exploded with delight, as if his horse had
just won the derby, and Mike was silenced for a
while. I remember my sense of immeasurable satisfaction as I walked out to my car that memorable
day, mumbling out loud to myself,

"Chalk-up another win for Dr. Kolmer."

~ * ~

We had many teachers in Dental School, each of whom contributed something special to our education. Of course we liked some, and felt some were unreasonable or difficult. Out of all those diverse people and even more varied subjects, probably our most singular experience was a brief episode with A.J. Donnelly, M.D., Professor of general pathology. Dr. Donnelly took the initiative to teach us something unusual. A convincing argument could be made that it was the most personally valuable single bit of information we received during those four years.

One day in general pathology laboratory, Dr. Donnelly said he was going to give us a treat before starting our regular microscope work. We all became alert knowing he had a good sense of humor. For example, he never raised his voice, but had deadly aim with chalk. If someone dozed off during his lecture, he unerringly would awaken him with a thrown piece of chalk. By the time the student had gathered his wits, Dr. Donnelly was facing the blackboard and carrying on as though nothing had happened, while the rest of us smiled and chuckled. The treat that day was a healthy lung suspended in a large bell jar. It was pink and lacy, floating in the preservative solution. He put the jar on a table, said nothing and gave us time to look it over. We had never seen a lung before, but were puzzled about the treat. Dr. Donnelly soon left the room, to return with a second bell jar. It contained an ugly mass of black tissue, no longer lacy but weighed down with disease, unable to float. The contrast was remarkable. At first he didn't say a word, while we digested the appearance of the two specimens. Finally Dr. Donnelly said the pink one was a normal lung, and the black lung was from an individual who smoked heavily for many years.

It was a sobering lesson. I looked down at the yellow stain on the two fingers I habitually used to hold my cigarette. Recalling how often I scrubbed those stains to remove the discoloration and lingering tobacco odor, I sadly wondered about the stains in my lungs. Silently I made a vow to quit smoking. That terribly damaged lung formed a lasting image, which helped me to break the vicious habit. In 1956 I smoked my last cigarette.

9.

ANECDOTES

The practice of endodontics can be very frustrating. This happens when root canals are effectively blocked with calcific deposits, forestalling the removal of the diseased pulp and elimination of infection. These calcified deposits are formed by the pulp in response to disease, or at times to aging. Unfortunately for the pulp's health, this process can be self-destructive: the calcifications tend to restrict the circulation, with diminished tissue viability and survival. The reduced circulation, in turn, also makes the pulpal tissue more vulnerable to infection. Of course, when the circulation is completely obstructed, the pulp dies, which usually in turn leads to inflammation of the tissues investing the tooth.

So treatment might be complicated when root canal therapy is prescribed for a tooth with canals filled with calcifications. Often the pulps in these teeth are both necrotic and infected. If these calcific deposits cannot be dislodged after lengthy tedious unsuccessful efforts to allow the debridement of the canal(s), then I, predictably, become very frustrated. Additional obstacles to root canal treatment are the natural curvatures of most canals, while being located within roots that are commonly tipped in various directions as they enter the jaw. Nonetheless, as a rule, the complex architecture of the roots and canal systems usually don't prevent treatment. I am pleased to report that most endodontically-involved teeth are successfully treated, and not otherwise extracted.

For myself, frustration is stressful, and, I suspect, it is equally so for many of my colleagues. Stress takes its toll in emotional fatigue. I used to assume all dentists were stressed by their work as I was. Of course this assumption was predictably incorrect: some must find dentistry wearisome, but most probably do not. I was amazed to hear a dental friend say, his staff literally had to throw him out of the office, otherwise he wouldn't know when to stop working. It's marvelous how different we all are. Isn't it?

What comes to mind is an interesting case bearing on canal blockages, but with an unprecedented climax. The patient was a middle-aged woman, who had been a patient of mine repeatedly. We always got along well and enjoyed a friendly relationship. Remember, as an endodontist, I only performed root canal therapy. She was determined to preserve her teeth, and had undergone extensive reconstructive dentistry with multiple fixed bridges.

On this visit she had a difficult root canal problem in the upper left second molar, which served strategically as the terminal support for an extensive fixed bridge—a method of replacing missing teeth. In this situation, the missing tooth was connected to crowns on the symptomatic second molar and the two premolars on the other side of the space. At this stage of dental treatment, the bridges were held on with temporary cement, for periodic removal to allow inspection of the teeth and soft tissues. Therefore, the bridge was readily removed and the second molar radiographed. My examination found dense calcifications in all the canals of this multi-rooted tooth. Moreover, the tooth's location far back in her mouth provided little space for examination and even more doubtful room for root canal instrumentation.

With so many obstacles to treatment, I felt obligated to share the negative prognosis with the patient. To my surprise, she insisted I try to treat it. Furthermore, I mentioned that there would necessarily have to be a charge for my time, even if treatment was unsuccessful. Undaunted, she was adamant I proceed, adding that she couldn't give up the tooth until she knew I had tried to save it. Finally, very emphatically she exclaimed, "I know if you can't succeed, no one else could."

Flattered into submission, I agreed to her demands.

After a prolonged difficult session, I finally had to conclude that all three of the canals were blocked with unpenetratable calcifications. The tooth was untreatable. Her bridge was replaced with temporary cement.

With my voice resonant with unconcealed regret, I told her of the disappointing outcome. She took it well, and seemed unfazed by the tedious efforts to stretch her mouth to its limits, while I had struggled to find room for my instruments. I was frustrated by the experience, felt emotionally drained, and remained slumped on my low stool. She got up, straightened herself, thanked me for trying, and started to leave. At the door to the treatment room she stopped, turned, possibly to say goodbye again, but hesitated while silently looking at me. Then to my amazement, she walked back, put an arm around my shoulder and said,

"Doctor Selden, it's all right. You did your best."

With a warm smile illuminating her face she turned and left. I'll never forget her gesture and words. As the professional, it was my role to comfort patients, not the other way around. This was the first, and only, time the tables were turned and a patient offered me emotional support. Though she was burdened with her own disappointment, she still made the effort to console me. A remarkable woman.

She'll never know what her few words meant to me. After all these years, I never forgot them.

~ * ~

The following story will seem remote from dentistry, but as it enfolds the connection will become clear. When I was a child the family spent summers out in the country in rural New Jersey, in a rustic cottage built by my grandfather. The amenities were primitive: backyard outhouse and water from a hand-pumped well. I thrived in this rugged environment, and eagerly looked forward to the summers. The expanded freedom it offered to this restless boy in his first decade was matchless, leaving an indelible imprint.

On one side of our wooded property, across a barbed wire fence, stretched endless acres of towering corn fields, part of Mr. Gilham's dairy farm. Along with a few friends we often wandered onto the farm, never missing a chance to watch the cows being milked. It was a memorable scene for a city-born kid, which I remember vividly. The dominant smell was cow manure; mixed with the sweetish scent of silage (a fermented, moist rich cow feed made from finely chopped field corn and other grasses); along with a tinge of warm milk. Farm hands silently stripped the milk with both hands, while seated on low stools next to the cows, their left shoulders braced against the animals hind quarters with heads turned forward. The sounds of the milk squirting into the pails were curiously soothing, like the patter of rain. We were entranced and, as if observing some solemn ritual, quietly watched. Milking machines had been marketed in the United States since 1918, I later learned, but the difficult economic times during the 1930s apparently placed the cost of such machinery beyond Mr. Gilham's reach.

Our small group wandered endlessly. There was something special about walking across open fields with no clear objective in mind, focused on the distant woods and far off hills. I can't recapture the actual emotions of the time, but do recall the sense of well-being and anticipation of what might lay beyond. It was an idyllic pastoral world, where thickly wooded rolling hills were spaced by fields carved out for farms long ago. Small wild crabapple trees flourished throughout the woods, or as solitary survivors in a clearing. As if generously offering their fruit, low reaching gnarled branches offered an easy climb.

Despite the threat of a belly ache from eating those hard, tart apples, I often ate them to quench my thirst and fortunately avoided ill effects. They didn't taste very good, but were a readily available treat, free for the taking. Inevitably we would end up back at Gilham's farmstead, in his apple orchard on a hill above his home. The apples we picked fresh from Gilham's trees were incomparable: crisp, juicy, and sweet. One day unexpectedly, after many such prior foraging ventures, someone mentioned that taking apples seemed wrong. He went on to say we should ask Mr. Gilham's permission. I guess our collective sense of guilt prevailed, because we all agreed. Without hesitation we marched down the hill and knocked on the farmhouse door. We had never seen Mr. Gilham before, so when he opened the door, his towering presence was more than intimidating, he was frightening looking. He stood over six feet in height, with a large muscular body, but his facial appearance was fearsome: his features were overly large and course; his lower jaw was huge and protrusive; the teeth seemed oddly out of position with prominently enlarged lips; and his skin was dark, thick and in folds. I couldn't help notice how his arms hung passively with the biggest hands I ever saw. When he spoke it was with a deep, husky voice resonating in a most unnatural way.

Mr. Gilham disappointed us. He was not disposed to say yes to our request to pick apples. He told us that since some boys had stolen apples, he would have to say no to us. Of course we were sympathetic, professed innocence, and assured him that we would never steal apples. True to our word, we never did again. Contrary to his scary appearance and voice, his manner was calm and he delivered his short speech slowly without emotion.

Years passed, and I don't think I ever thought of Mr. Gilham again, until one day while a student in dental school. I was reading the chapter on diseases of the endocrine (hormonal) system, when I came across a photograph of a patient with advanced Acromegaly. I was startled and shocked. There was Mr. Gilham looking back at me. I learned that all his physical changes and disfiguration were due to the excessive secretion of the hormone from the anterior pituitary gland, nearly always due to a tumor. The hormone abnormally stimulated the boney sutures in the face and head to grow new bone, creating the typically grotesque appearance, referred to as "lion-like." The hands and feet were also affected. Today both surgical and medical approaches to therapy are available.

The pituitary gland certainly lives up to its reputation as the master gland, governing and coordinating many bodily functions. The pituitary not only produces the powerful growth hormone discussed above, but also the opiate-like compound called endorphin, released during acupuncture.

10.

REALITIES AND ILLUSIONS

Despite the extraordinary progress in the science and technology of endodontics, the diagnosis of pulpal disease can occasionally be troublesome. Diagnostic problems arise when the patient's description of their discomfort is vague, lacking specificity, and the usual examination procedures, including radiographs, are inconclusive: fails to locate the source of the pain.

Most dental patients realize that radiographs are important. They are actually more than important. Radiographs are truly irreplaceable, and are essential to a thorough endodontic examination. But as part of the work-up, they have well-known limitations, and their value depends on skilled interpretation.

The following case reports will illuminate the subject:

#1. The patient complained of pain in the maxillary right anterior region. The radiograph showed a very large area of radiolucency (bone loss) over the apex of the right lateral incisor. It also appeared as though the roots of the teeth in the area were displaced, presumably by the pressure of the disease process. The findings were typical of an aggressive infection and, with the lateral in the middle of the bone breakdown, it seemed the likely cause. Despite the radiographic evidence, the examination was incomplete. Corroboration of my preliminary diagnosis came next with tests of all the area teeth for pulpal reactivity to temperature challenges and electric stimulation. These tests provide valuable indications of pulpal health or disease. Keep in mind that if the infection was coming from the lateral, its pulp would be dead, and therefore unresponsive to stimuli.

Applying ice to teeth is a common low-tech temperature test. It can serve as a general screening, or even strategically help solve a diagnostic dilemma. When a normal pulp is tested with ice, the patient usually can detect the cold, but doesn't experience pain. A dead pulp is on the other end of the spectrum: it can't feel anything when ice tested. Sometimes pulps are alive but irritable. In an early stage of pulpal disease, ice could cause sharp, lingering pain. Curiously, as pulpal disease worsens, sensitivity to cold disappears, only to be replaced by pain to heat. At this stage, the heat induced pain can be relieved by cold. These changing patterns of responses to temperature stimulations provide useful tests to help with the diagnosis of pulpal diseases.

Once in a while, knowledge of these temperature reactions allows a dramatic snap diagnosis, before examining the patient. In the progressions of pulpal diseases from health to necrosis, the time intervals between stages of pulpal deterioration varies widely in each instance, and symptoms are unpredictable. However, there is one stage, described as an acute suppurative pulpitis, where heat intensifies the pain. If by chance, this stage of disease lingers, the patient is assaulted with an unrelenting severe toothache. At its worst, the normal warmth of the mouth is enough to generate this agonizing condition. The snap diagnosis is made when a patient with a thermos of ice water is greeted. Many of these sufferers fortunately discover that cold will stop the pain, albeit only momentarily, so they must continuously sip the soothing cold water. Since no other oral disease creates this unique acute affliction, the diagnosis is obvious: it is an acute suppurative pulpitis. Root canal treatment will stop the pain and save the tooth.

To return to case #1: I applied a small piece of ice first to the lateral, and then to the teeth on either side. To my surprise, the patient said they felt cold in the lateral and the cuspid, but not in the central incisor. This response cast serious doubt on the accuracy of my initial radiographic-based impression. Yet, I speculated, ice testing could be misleading.

So, next I used the sophisticated device called a pulp tester. It utilizes an electric current to stimulate the sensory nerves of the pulp. The test is performed on the tooth surface, one tooth at a time. The patient is told they will feel nothing at first, but if a tingling sensation is detected they should raise their hand. At which point the test will immediately be stopped. As with ice, if the pulp were dead, then no sensation would be detected, even with the pulp tester at maximum intensity. Surprisingly, the suspected lateral felt the electric stimulation at a low level, indicating a normal pulp. When the central incisor was tested, it failed to respond to the highest setting of the pulp tester. With the ice test and pulp test in agreement, root canal treatment on the central was indicated. The finding of an infected necrotic pulp in the central confirmed the diagnosis. Rapid healing of the damaged bone followed.

The original radiograph created the illusion of lateral involvement. Radiographs are two dimensional views, and thus could not display the relative depth positions of the roots of the lateral and central within the jaw bone. In this instance, the lateral root was actually deep in the jaw behind the root of the central. Therefore the osseous destruction, seen on the radiograph, caused by the root canal infection draining from the central into the bone, was actually in front of the lateral, not involving it. This is a good example where a hasty diagnosis only based on radiographic findings, could result in a mistaken treatment.

~ * ~

#2. This patient had a failing endodontically treated upper lateral incisor. The radiograph showed a significant radiolucency around the apex, that looked like a chronic infection. In this case the prior root canal treatment appeared satisfactory, and the pathologic apex was readily accessible surgically. Therefore surgery was performed, wherein the pathologic tissue around the apex was removed and the opening at the end of the root resealed. Despite normal soft tissue healing after surgery, the patient continued to report tooth tenderness, and the apical radiolucency persisted. An important sign of treatment success is the re-growth of bone around the apex, especially after surgery. In this situation it did not occur. Eventually the tooth was removed, and revealed the undetected cause of the failure: the root itself had a major crack from mid-root to the apex. The radiograph failed to show the vertical root fracture, since it was obscured by the dense radiopaque root canal filling. I was at first surprised I hadn't spotted the crack during surgery. But, on further review, I realized the crack was hidden on the back of the root where I couldn't see it. There are situations where the most expert treatment will fail, due to uncontrollable factors.

~ * ~

#3. The limitations of the radiograph are important to recognize, although we depend on them to provide essential images of the teeth and surrounding structures, otherwise clinically invisible.

Root canal treatment was performed on the patient's upper left second premolar. Therapy was uneventful, and the single canal found within the tooth conformed to the usual configuration. Six months later the patient was seen with the complaint of tenderness to biting on this tooth, and soreness to finger pressure on the gum in the fold above the tooth. A radiograph showed the formation of a small radiolucency at the apex, not seen previously. Puzzled at the cause of the deterioration, I decided to surgically explore the area. With the root tip positioned near the jaw surface, just under the gum, surgical access was uncomplicated. Routinely in situations like this, the end of the root is removed—called an apicoectomy. When I cut off the root end, I was shocked to find that behind the main root, a small second root branched backwards near the apex. This second apical root was not seen on the radiograph, and was also hidden from my view surgically until I held the amputated apex in my hand. If I had chosen to re-treat the root canal instead of performing the surgery, treatment would have failed again. It would have been mechanically impossible to clean and disinfect the canal of this small root end branch. Therefore, the ongoing seepage of sepsis into the jaw from this very rare root end formation, would have been enough to perpetuate a chronic infection. Following surgery, healing progressed normally and symptoms never returned. This case illustrates how limited radiographs are at times.

~ * ~

#4. I can't help thinking of this final case as particularly bizarre.

The twenty-five-year old male patient recalled a blow to his upper left permanent central incisor at a very early age. He couldn't remember exactly when, except that he was in grade school. The tooth recently became symptomatic for the first time. I wasn't surprised with the radiographic image of the tooth: it was shorter than usual, with a blunted shape; showed an extremely wide, irregularly formed canal space (described as "blunderbuss"), instead of the normal narrow tapering canal. The radiograph allowed me to closely calculate his age at the time of injury to the tooth. It isn't a secret that the maxillary permanent central erupts into the mouth around the age of seven. What isn't generally known is that it takes three more years for the root to form. At seven only the crown of the tooth is fully developed, while the root appears as a short thin-walled extension from the base of the crown. The biological growth of a mature root is an incredibly complex process, but it depends completely on the primitive pulp. Root growth is halted if the crown sustains a blow of sufficient force to kill the pulp during this immature period. Therefore, a fairly precise determination of the patient's age at the time of the trauma can be made. It is based on how much root had formed before the pulp died. In this case, I correctly judged he was between eight and nine years old when the tooth was injured.

For such undeveloped teeth root canal treatment is usually prolonged: All remnants of necrotic debris must be gently removed from the very large, thin-walled, canal space; infection eliminated; and special dressings used to stimulate repair over an extended period of time.

When I judged conditions were right in this case, the canal was permanently filled with gutta-percha (an isomer of rubber, basically inert, does not stimulate a foreign body reaction in the tissues, is readily softened by heat to allow close adaptation to the canal walls, and is still in use). The tight filling of root canals is the final step in treatment. This is essential to prevent oral sepsis from re-contaminating the canal and spreading infection into the jaw bone.

When the patient was seen again some months later for a routine postoperative examination, he reported that the tooth had been persistently tender. A new radiograph failed to show any changes. The chronic soreness was troubling. With the patient's agreement, a surgical exploratory procedure was performed.

The elevation of a gum flap for visual access to the tooth, disclosed a startling reality. Instead of finding an intact root, I was shocked to see a large bulk of the solid gutta-percha root canal filling protruding from the center of the root. The radiographs taken during treatment gave no indication of this pathologic root opening, through which the filling material extruded. Under the circumstances, it created the false radiographic image that the material was appropriately contained within the tooth. Actually the gutta percha was out of the canal, resting on the front of the root, giving the radiographic illusion of being in the canal.

I was astonished by what I saw. Though momentarily frozen with inaction, I quickly recovered and performed the necessary corrections: the excess gutta-percha was removed, the area debrided, and the flap replaced. The healing was normal and the patient had the use of this unusual tooth for many years.

This case was a sobering experience: I still find it hard to understand how I could have been unaware of the opening in the root—a most likely result of the early trauma. The painful lesson taught me to increase my wariness with radiographs, especially in complex cases. Now, you the reader, have also learned that radiographs can portray illusions, at times remote from reality.

Though most root canal diseases are successfully managed non-surgically (the canal is treated through an opening in the crown), the need for an occasional surgical approach still exists. Note how cases #3 and #4 healed only after surgical exposure of the roots revealed the nature of the defects, and made possible their correction.

11.

PARADOX

Even though by now you must be concerned with the issues of interpretation of radiographs, don't get discouraged. The reality is that radiographs are very much a paradox: Yes, dental radiographs are always essential; yet occasionally could be misleading. It is a potential diagnostic enigma. It helps, when possible, to avoid making a diagnosis based on radiographs alone. To further challenge dental radiographic interpretation, there are periradicular bone scars. They present truly ambiguous radiographic images.

Although the reliability of nonsurgical root canal therapy has been proven (in the range of ninety percent), periodic clinical follow-ups with radiographic examinations are advised. The radiolucencies commonly seen around the apecies of infected canals are often signs of inflammatory breakdown of bone. After root canal treatment the radiolucency should fade if healing occurs. A standardized radiographic technique allows periodic views of these areas to assess healing. If the radiolucency doesn't disappear, or it shrinks but remains evident, then a diagnostic dilemma exists. In evaluating the response of root canal treatment, especially in cases lacking associated clinical symptoms, the radiograph often assumes critical importance. Here are typical cases.

~ * ~

#1. An otherwise healthy male in his late thirties presented with a painful upper left lateral (strictly coincidental how many of the cases in this book are upper laterals). The tooth was tender to percussion and showed an apical radiolucency. Ice application and electric pulp tests failed to elicit any response from the tooth. With all findings in agreement, root canal therapy was performed. The diagnosis was confirmed by the necrotic infected pulp found within the canal. The tooth responded well to treatment but the patient was lost to regular follow-up. He returned seven years later. He had moved out of the area and was now visiting, and thought he would have the tooth checked. When I asked how the tooth felt, he said it hadn't given him any trouble. I was pleased to see him again, and ordered a radiograph of the lateral. Despite the lack of any symptoms, the new radiograph showed a deficiency in healing of the radiolucency. Together we studied the new radiograph and compared it with the seven year old pretreatment film. There was no denying that the radiolucency was still evident. If healing had been complete, as it should have been after seven years, the radiolucency would have disappeared. The patient was unsettled by this failure of bone repair, and opted for the surgical excision of the area with submission of the tissue for histopathological examination. During the surgery an apicoectomy was performed along with the removal of the attached tissue from within the radiolucency. To insure a tightly sealed root canal, an apical filling was placed. He healed uneventfully. But to our surprise the pathology report unequivocally described the lesion as a scar, with no signs of inflammation (the finding of inflammation is a reliable indication of infection.) This was a rare finding. The scar is healthy repair tissue, only without bone. But regretably, scars look very much like an inflammatory radiolucency on a radiograph. The patient was relieved with the scar diagnosis, but I felt uneasy. I pondered, wasn't there some way to identify a scar on the radiograph? I had to conclude there usually wasn't. Surgery had to be considered when there was doubt as to the nature of the radiolucency.

This showed how scar tissue repair could create a diagnostic problem. Why bone didn't form is unknown. Fortunately, most inflammatory radiolucencies heal with bone formation after nonsurgical root canal treatment.

~ * ~

#2. This case is different. It dealt with scar formation after surgery. The adult female patient was referred following recent surgical removal of a very large cyst from the right anterior maxilla (upper jaw). Examination confirmed that the pulps in the upper right cuspid, lateral and central incisor teeth—whose apecies were enveloped by the postsurgical bony defect—had necrotic pulps. If root canal treatment wasn't rendered, the inevitable sepsis from these teeth would contaminate the surgical wound, prevent healing, and create a potentially large dangerous inflammatory lesion. After root canal therapy on all three teeth I followed healing for two years. I was fascinated watching this huge radiolucency shrink as recalcification began peripherally. Eventually new bone covered all three apecies—a confirmation of successful root canal treatments. Bone cannot swell in the presence of inflammation, as does soft tissue: Bone breaks down in the presence of inflammation. The body's efforts at bone repair can't begin until all irritants are removed. Therefore the bone deposition around the teeth was reassuring.

Probably because of the large dimensions of the bony defect after the cyst removal, total calcification of the defect didn't occur. Instead a small bone scar formed in the center of the defect. The nature of this residual radiolucency was not a diagnostic problem. First of all, its isolation in the maxilla, not in contact with or even near the apecies of the three endodontically treated teeth, eliminated a possible new inflammatory lucency. Secondly, it is not unusual for scar tissue to form along with the calcified matrix of bone repair. Why the central area of the defect did not totally calcify is unknown. Of course radiographic follow-up examinations will confirm the benign and stable nature of the residual radiolucency.

~ * ~

#3. This is a more complicated case. There are multiple teeth with prior root canal treatment, along with a history of apical surgery on one of them. The immediate concern was a symptomatic upper left central incisor. All tests confirmed the tooth contained an infected pulp, and root canal therapy was promptly instituted. The radiograph of this tooth also showed the adjacent lateral and cuspid with root canal fillings and troubling findings. The left lateral incisor showed a well-defined, markedly radiolucent area on the apex, a shortened root, a possibly leaking root canal filling, and a history of multiple apicoectomies. Nonsurgical retreatment of the lateral canal was performed. Follow-up radiographic observation was particularly advised, to confirm whether the suspicion of postsurgical scar was indeed correct. The cuspid was asymptomatic and showed a normal apex.

Five years later the patient returned with minor symptoms in the same area. A new radiograph showed the radiolucency on the lateral was unchanged, while the apex of the cuspid had deteriorated with a small radiolucency now present. Nonsurgical retreatment of the cuspid was advised, which the patient chose not to treat at the time. When seen again fifteen months later, a sinus tract (a pathologic path for drainage of pus) was present on the gum over the lateral incisor. A new radiograph provided no help in determining which tooth was the souce of the drainage, the lateral or cuspid. The uncertainty was readily solved by passing a thin, soft silver wire into the opening of the sinus tract and moving it to the origin of the pus. A radiograph with the wire in place graphically showed the drainage was coming from the apex of the cuspid. With the patient's approval, nonsurgical retreatment of the cuspid root canal was finally performed. Within days the drainage ceased and the sinus tract healed.

Two years later, the symptomless patient was reexamined. The new radiograph showed healed apecies on the cuspid and central. In addition, the compact bone over the lateral apex had reformed, separating the tooth from the peripheral persistent radiolucency. When the patient was last seen after eight years of observation, all teeth were clinically and radiographically healthy. Of special interest was the apical lucency near the lateral: it was unchanged in size and no longer seemingly connected to the lateral apex. Without a doubt the early diagnosis of a periradicular scar following surgery could now be unequivocally confirmed.

12.

NIGHTMARES

Dreaming is common, and nightmares are not rare. What I think is rare is repeatedly experiencing the same nightmare, as I did for more than twenty years after graduation from dental school. It took me a long time to understand why this replay of a dental school theme was so persistent.

To make sense of this sorry tale, I'll have to sketch some background. An apt description of dental school, in the early 1950's, would be to call it an elegant pressure cooker. The fact that we were expected to absorb an incredible amount of scientific material is not surprising. For instance, in the junior year, we took eleven separate final exams. This created considerable excessive stress. The pressure was intense since the final exam was the only test for some courses. We would pass or fail based on one examination. The threat of expulsion for even one course failure was probably more fiction than reality, yet we took it seriously. If that wasn't enough, we were also responsible for completing many time-consuming laboratory technique projects.

And of course the junior year was special. It was pivotal: we started treating patients for the first time. To admit we were nervous was an understatement. In addition, we were confronted with the long-established dental school practice of clinical requirements. These were rigid standards of minimal numbers of completed procedures in all phases of restorative and prosthetic dentistry. The avowed purpose was to ensure students were clinically qualified to graduate. This sounded reasonable, but the negative effects were significant. To all of us neophyte dentists, the requirement scheme was controversial and upsetting, raising the temperature of the pressure cooker. Many of us spent considerable time in heated discussions criticizing the rules, over endless cups of coffee. We felt the emphasis was inappropriately placed on numbers, instead of quality. There was no recognition that some talented students could demonstrate skill with a few restorations, while others might need many repetitions. Moreover, fulfilling requirements replaced all other concerns. Numbers alone were paramount. The serious intent to hold us to these accomplishments was never in question: we knew graduation was delayed until all requirements were satisfied. Dominating one wall in the clinic was a huge, mural-sized, mounted chart where each student's completed units were checked off. The chart was a display of wrongheaded priorities on public view. Strangely, it reminded me of the wooden stocks mounted in town squares for discipline of offenders during early times. Both the requirement chart and the stocks were tangible intimidations. The message was clear: follow the rules and avoid punishment. Regrettably, nowhere in this system was there a provision for recognition of excellence. Somehow the authorities had also decided to place the total responsibility for obtaining patients on the students, not the school. A shortage of patients was an unacceptable excuse for not fulfilling the requirements.

Our frustrations about dental education led to comparison with medical school, only a block away. During the first two years, the programs for medical and dental students were similar. But when we both started treating patients in the junior year, a dramatic difference evolved. Medical students had experienced physicians as mentors to emulate, and nothing like requirements existed. For dental students, the junior year was the beginning of a very trying period. Unlike medical students, we sadly had an adversarial relationship with many of the clinical dental staff. Nor did we often have the benefit of assisting experienced dentists in treating patients. We learned on our own; and relied on upper classmen for advice. Instead of providing emotional support, some of the staff at times were aggressively antagonistic. Going through dental school had echoes of the hazing rituals of college fraternities: you could refuse to participate, but then you couldn't join the club. Our repressed anger heightened unpleasant memories.

My response to the intense pressure of completing the requirements was never to walk if I could run, and work like a demon. As a result, I readily fulfilled and even exceeded the minimum requirements during both clinical years. This was especially true in prosthetics, where I additionally constructed new dentures for friends of the head of the department—his colleagues from the undergraduate school. I was flattered to be asked, and somehow had no trouble fitting the extra work into my hectic schedule. Bear in mind, that I not only took the necessary oral impressions and measurements of the patients for new dentures, but also actually performed the entire fabrication of them in the laboratory.

Now, we come to my nightmares. I think the cumulative stress and anxiety of dental school contributed to these nightmares. They probably could be likened to a form of post-traumatic shock. The nightmares were always the same, and began shortly after graduation:

"I found myself in the dental clinic looking up at the requirement chart. Painfully I realized I hadn't fulfilled the requirements. The school year was coming to an end and I was paralyzed with despair."

When the dream gratefully came to an abrupt end, I awoke, but was left with a residue of depression, puzzled about the dream. After all, in reality, I never had any trouble meeting the requirements, even exceeding the minimums. Why this nightmare? It took many years for me to realize the obvious. I certainly had been deeply worried and anxious about the requirements as a student, which left a powerful subconscious record. Therefore, through the years, when I was worried or deeply concerned about something, I could reliably summon this dream to express those profound emotions of anxiety while asleep. During my wakeful hours, those feelings were submerged. Clearly those anxieties needed to be dealt with and the nightmare was a package I could depend on. Fortunately, I eventually outgrew the need for this nightmare, and moved on to more pleasant dreams. Wonder why it took so long for me to figure this all out?

13.

PRAGMATISM AND SCIENCE

By the senior year, I had gained enough experience treating patients to concentrate on increasing my speed. This was important, since dentists were only paid for completed treatment. It dawned on me, that this fee for service tradition could be the basis for the rigid numerical clinical requirements. After all, if that is how we were to earn a living, then students had better get used to the idea that production was essential. So with encouragement and support of the faculty, I honed my skills to be able to rapidly "drill and fill" in one visit. The old cliché "time was money" was our guiding mantra. Speed would pay off.

Therefore, we were shocked to learn how some older dentists intentionally prolonged cavity treatment. Instead of cleaning out all the decay and permanently filling the cavity in the same visit, they partially removed decay and placed temporary fillings. In this way their income increased by the additional number of visits involved in completing a permanent restoration, since a fee was charged for each cavity treatment visit. We only gave passing notice to the universal use of zinc oxide-eugenol cements for the temporary fillings. As discussed earlier, it was common knowledge that eugenol would reduce or eliminate pulpal pain. On the other hand, eugenol's ability to kill decay bacteria was likely overlooked or unknown. Since removing decay in stages evolved as a purely pragmatic economic consideration, I was pleasantly surprised to eventually learn how it worked to the mutual benefit of patient and dentist.

By the way, the routine use of eugenol containing temporary fillings had a distinctive side effect: it left a pungent odor of cloves (eugenol is the essential constituent of clove oil) in dental offices, as well as lingered on dentist's hands. Consequently, generations of people came to associate the smell of cloves with dentists. I guess it could have been worse.

I understood this practice began during the economic depression of the 1930's. Dentists suffered fiscal hardship along with everyone else, and undoubtedly were motivated by a need to maintain their income. They also had to lower their fees during those difficult times, and worked longer hours in an attempt to compensate. On one hand we were sympathetic to the economic realities of the past, but on the other hand, as youthful idealists, we couldn't help view this practice as crass commercialism.

By the late 1950's research into pulpal biology made considerable progress, much to the credit of Doctor Harold R. Stanley. His work was of particular interest to endodontists. With our focus on the health and disease of the dental pulp, these new findings provided an improved scientific basis for diagnosis and treatment.

Amazingly, research showed that the removal of decay in stages produced minimal pulpal irritation. This allowed the pulp to respond with the maximum deposition of a calcified barrier under the approaching decay. These protective calcifications, called re-parative dentin, are the pulp's response to the inva-sion of decay. However, unless all the actual decay is removed in a timely fashion, decay could win the race of penetration into the pulp, with painful conse-quences. The inevitable complication of untreated pulpal infection is the spread of infection into the jaw bone itself.

It was enlightening to learn that when a layer of decay is removed, and a zinc oxide-eugenol cement is placed, the metabolism of the remaining decay bacteria is interfered with: The bacteria's food source is eliminated; and the production of lactic acid by the bacteria is diminished or halted. This acid decalcifies the dentin ahead of the bacteria and is very toxic to the pulp. So the temporary fillings were beneficial by favorably shifting the balance towards control of the decay and survival of the pulp.

On the other hand, when decay is removed all at once the pulp doesn't always respond favorably, re-sulting in chronic inflammatory changes. The deeper the decay, the more likely these adverse pulpal ef-fects.

An important lesson has been illustrated. Sometimes the history of dentistry, and for that matter also medicine, has been characterized by the successful clinical management of disease prior to an understanding of the mechanisms involved. Thus in lieu of sufficient scientific and biological guidelines, treatment methods will continue their pragmatic and empirical evolution.

14.

ARMY DENTISTRY

If Dental School germinated dentists, then my two years of active duty service in the U.S. Army Dental Corps, following graduation, matured me as a dentist. Professional development is a slow process. It takes time (think years) to acquire experience, hone skills, develop critically well-balanced judgment, and build self-confidence.

When I taught endodontic graduate students at Temple Dental School in Philadelphia I predicted,

"For the first five years after graduation you will rely on what you've learned; during the next five years you will start to question the dogma you relied on, and will begin to think creatively. Our profession's growth relies on each generation's bold new

ideas. Some of you can become the next intellectual leaders."

~ * ~

For the most part, army basic training is tough grunt work turning civilians into soldiers. For dentists, physicians, veterinarians, and nurses, all of whom are given officer commissions from day one, military basic training is a walk in the sun. Back in the mid-fifties, at the Army's Medical Field Service School, Fort Sam Houston, located in San Antonio, Texas, we played at being soldiers for five weeks.

But the training sessions on a likely nuclear war, confounded and frightened us. Those were sobering days! The prospect of nuclear devastation, with the potential for awesome numbers of fatalities and casualties was daunting. Initially, my skills and training in dentistry seemed unequal to the task. In time, I came to realize that I was an essential part of an extremely well organized and coordinated team of health providers, where my special talents were important. We all trained together: observing, for example, a mock war that showed how corpsmen quickly brought casualties back to the combat area medical facility, the MASH unit (Mobile Army Surgical Hospital). These hospitals created in 1945, were conceived by Doctor Michael E. DeBakey and others, while gaining fame through popular TV programs. The MASH units served effectively through the conflicts in Korea, Vietnam, and the Gulf Wars, and were finally deactivated by the Army in 2006; replaced by medical units called Combat Support Hospitals. It became clear, that regardless of our training—medical, dental, veterinarian, or nurse—we were expected to work together, and to step up to replace each other if the need developed. My self-image as a dentist was substantially elevated. I foresaw where my training in oral surgery, medical evaluation of patients, pharmacology and drug administration, prepared me for an expanded role in a MASH unit. Fortunately, I wasn't called on to serve during combat, but worked in a stateside dental facility.

~ * ~

My wife Tamara, baby David and I drove across the country in our new, bottom-of-the-line, stripped-down Plymouth along route 40—the major East-West highway prior to the construction of the Interstates—from New Jersey to my duty station at Fort Riley, Kansas. The assignment to Kansas rankled mightily, since I had listed Hawaii, Cuba (we were still friends), and Germany as my choices on the preference form the Army provided us with at Fort Sam Houston. The middle of Kansas was the last place I wanted to be. Of course, I had heard that Fort Leonard Wood was the worst, so I should have been grateful. I was learning that the Army's decisions were based on their needs, not mine. The preference form was obviously a charade.

The further West we got, the hotter it became. Maybe the people in the Midwest were accustomed to the heat, and the periodic droughts (that summer marked the fifth year of a severe drought) but we certainly weren't. By the time we crossed the Mississippi river, it was obvious the lush green of the East was replaced with a brown dried-up landscape. Many rivers were now totally dry, with only their names forlornly posted on the bridges over which we drove. And in our non-air-conditioned car, with temperatures hovering around 100 F, baby David developed a world class case of diaper rash.

That summer of 1955 the 10th Infantry Division at Fort Riley was training for replacement of "The Big Red One," which was finally coming home after serving from their initial landing in North Africa in 1942 to German occupation. I soon learned that this transfer of divisions was known as a "Gyro-scoping" maneuver. With all 10th accommodations, on and off post, committed to the 1st, housing was almost unavailable. After a day of futile searching I booked a room in a motel a few miles from the post, and after dinner went out determined to find a place to live. To my delight, when I called on the manager of a garden apartment complex expecting to list my name for an opening, he told me someone just moved out. I took it sight unseen. The apartment was just fine. Tamara was much relieved with my success, and its location in Junction City, the town just south of the Fort was a short daily commute.

I was initially assigned to Forsyth the smallest of the dental clinics—the Funston area had the largest clinic while the hospital's dental clinic also provided

an oral surgical capability. Forsyth lent to easy socia-
lization between us new young dentists and a jolly
middle aged Light Colonel dentist in charge, who did
no work, and passed his time flirting with the WAC
dental assistants. The place is memorable because it
was where I started drinking black coffee. Until then,
I always used milk or cream to soften the coffee taste
and change its appearance to an appetizing mocha
color. A small closet-sized room in the wooden clinic
was designated the officer's lounge. Of course it was
as barren as the rest of the place, except for a pot of
coffee kept warm on an electric hot plate on a small
shelf, along with cups, sugar jar, and an open can of
condensed milk. Some time midmorning on my first
day, while waiting for a patient to get numb after I
had injected novocaine, I went for a cup of coffee. I
found what looked like a clean cup and poured in
the coffee and then a generous portion of condensed
milk. The taste of the condensed milk was strong
and new to me. I wasn't sure I liked it, and thought
with time I'll get accustomed to it. But before I swallowed my first mouthful I felt a rather large object in
my mouth. When I spit out the coffee I was appalled

to see a live cockroach dashing away. It must have crawled into the open can of milk. Still tasting the condensed milk, I knew I would never use it again. As time went on I got to like the strong taste of black coffee, and to this day I still drink it that way.

Which brings to mind the time I was on emergency call. The forty-some-odd dentists took turns with dental emergencies after hours. We could go home, or for singles, return to their BOQs (bachelor officer quarters), and be available if needed. Once, I received a call about 1:00AM from the base hospital, that a soldier had turned up in severe pain from what appeared to be a toothache. Since I had previously spent a few months working in the hospital dental clinic, I knew my way around. But it was the first time I would be alone with a patient. Usually, by the time I arrived in the morning, the clinic was open and functioning; the dental assistant was there to bring the necessary supplies, such as local anesthetic and syringe; conduct the patient to the radiology room for dental x-ray pictures; and provide whatever assistance was needed. As I conducted the sergeant down the hall to the clinic, my mind was trying to recall where everything was, including the room's light switch. It took time to locate the electric power switches for the dental units, find the switches for the central suction pump and even locate the valve for water. The suffering patient silently sat in a dental chair while I wandered around like a blind man checking for the hidden switches. I imagined he was wondering what kind of a dentist it was his bad luck to draw. Well, eventually I had the place turned on. Confirming the source of his pain was easy: he pointed to a badly decayed lower molar. The x-ray of the tooth went smoothly, but developing the film took time with the old-fashioned liquid chemical solutions. I almost dropped the film while hand dipping it in the chemicals to hasten the process. The film finally cleared enough for me to see the extremely long, curved roots, surrounded by dense, thick mandibular jaw bone. Removing a lower molar was a challenge at best, and was complicated by the relative difficulty of obtaining an effective block with the local anesthetic. At times, a tooth like this must be sectioned in order to extract it. When coupled with dense bone, elevation of a surgical flap might be required to allow removal of some of the bone along with the roots. I had experience with such surgery, but worried about its feasibility without an assistant.

Despite my anxieties, the block worked perfectly, and I extracted the tooth in one piece without complication. The patient was grateful, and I felt much relieved.

The story has a sequel. Some months later, I was stopped by an M.P. on post for speeding and not stopping at stop signs. I knew I was guilty, so quickly started formulating a plausible excuse. When I turned to look up at the M.P. standing along side the car I was startled to recognize my emergency patient. My strategy immediately changed as I smiled up at him and asked,

"Didn't I come out in the middle of the night a few months back to remove an aching tooth for you?"

He looked more carefully at me, smiled recognition, and said,

"Captain, you sure did me a fine service. So please watch your driving in the future."

That was it. He saluted and left. I drove slowly the rest of the way to the dental clinic and mused about the wisdom in the old saying, "what goes around, sometimes comes around."

~ * ~

Those army years coincided with the height of the cold war. Strategic Air Command B-47 jet propelled bombers were constantly on patrol around the Soviet Union with their precious cargos of nuclear bombs. In contrast, at Fort Riley, Kansas, where I was stationed, the Army had helicopters and single engine Piper Cub reconnaissance planes. The contrast was stunning. The government's policy of "deterrence," which relied on SAC bombers, inescapably trumped all other budgetary needs, with the Air Force swallowing a lopsided huge portion of defense spending. As a result, we were unable to fill our dental supply needs in the Army. The situation became acute, especially with our current mission to prepare a replacement division. I couldn't help think, if we (dentists) were combat soldiers, we would be running out of bullets.

The SAC bases were more special than I realized when I decided to manage a trip back East on one of their planes. It was common knowledge that with a uniform and official leave papers the Air Force would be willing to give you a ride. I got in touch with a friendly Piper Cub pilot—also a regular patient of mine—for a lift to Schilling Air Force Base due West near Salina, Kansas. He was happy to arrange the flight since he had to accumulate hours of air time to maintain his qualifications. As we landed on Schilling's long runway in our Piper I couldn't help think we were like a little gnat, dwarfed by the immense facilities and huge bombers I saw parked around the field. When we finally coasted to a stop, I was startled out of my reverie by someone shouting that I should raise my hands. I looked down out of the cabin window and was shocked to see a military vehicle filled with armed soldiers pointing their guns at me. I opened the plane door, keep my arms raised, and asked them not to shoot since I was on their side. The officer in charge wasn't impressed and asked,

"What the hell are you doing landing on a SAC base? Didn't you know it was forbidden?"

I pleaded ignorance, and explained I was hoping to get a ride back East. The officer then turned his attention to my pilot and said,

"As soon as your passenger gets out, I want you to take off and fly the hell out of here."

He finished with the threat,

"If you ever try this stunt again I'll throw you in the brig."

I climbed out, and immediately, without so much as a passing goodbye wave by the pilot, the Piper spun around and was off.

SAC security was certainly impressive. The situation served to remind me how serious and dangerous were the times.

Eventually I did get a nonstop flight across the country to Massachusetts, but not in one of those sleek B-47s. My air taxi was a lumbering cargo plane, the C-119 (euphemistically called the "Flying Boxcar"). These unusual looking twin-boom transports were in continuous operation since 1947; finally replaced around 1960.

~ * ~

The Dental Detachment's responsibility was to examine every soldier in the new 10th, and restore their oral health, as needed, before they could be sent to Europe. I was shocked to discover a few soldiers with rampant decay and acutely inflamed gum conditions. Clearly, they never had any dental care and were totally negligent with oral hygiene. I couldn't understand how they passed their pre-induction physicals. Contributing to, and maybe explaining their dreadful dental health, many of these young men clearly, by their own admission, had a severe phobia—they were terrified of dentists. These dentally handicapped soldiers would never be sent overseas.

Now let's talk about *trench mouth,* a misnamed oral disease. The term originated during the First World War, when many soldiers developed an acutely painful, bleeding gum condition, while hunkering down for long periods of time in awful dirt trenches. The assumption that this was a highly contagious infection, having spread among the soldiers seemed plausible, but was misguided. Though terms for this oral disease are somewhat in flux, it commonly is called *Acute Necrotizing Ulcerative Gingivitis*: a good descriptive name. Now we know that though huge numbers of Fusospirochetal organisms (the main bacterial culprits) are found in oral smears, the disease is not readily transmittable: namely, it is not a contagious disease. The young men in the trenches were all run down, dealing with the same miserable conditions: dirt, body filth, exhaustion, inadequate diet, fear, lack of sleep, poor oral hygiene (if not totally lacking), smoking cigarettes, and if available, drinking alcohol. The combined effect of the above factors predisposed many of the soldiers to suppression of their immune systems, leading to an acute flare-up of *ANUG*. It is a very painful infection. Not surprisingly, some of those soldiers in the 10th, whose deplorable oral conditions I spoke about above, eventually developed *ANUG*.

The treatment for *ANUG* is very effective: there are local and supportive measures. Locally the mouth is gently cleansed, antiseptic mouth washes are used at frequent intervals during the early days, and if conditions are severe enough, oral antibiotics are helpful. Eventually a thorough dental cleaning removes all accretions. Supportive measures are equally important: cessation of tobacco smoking and alcohol ingestion; a soft but well balanced nutritional diet along with vitamin supplementation especially C and B complex; and finally the patient is advised to obtain adequate rest, both physically and emotionally, and drink plenty of liquids. The prognosis is favorable, as long as the patient is cooperative.

~ * ~

Fate worked in our behalf one day when my wife and I along with infant son David decided to visit my friend and classmate, Paul Weiner, at Forbes Air Force base (one of the many SAC bases) outside of Topeka, Kansas. The sixty miles east on route 40 to Topeka was an easy straight run. While our wives socialized, Paul and I toured his dental facilities. Coming from Fort Riley's dental clinics in old wooden barracks, with outdated dental units and supply shortages, his clinic, in a brick building, with new state-of-the-art equipment, was staggering. I couldn't help mumble,

"Why didn't I have the good fortune to be assigned to the Air Force instead of the Army?"

Paul grinned with delight, and then led me to an adjacent brick building while saying,

"Now wait till you see this."

What I saw in this other building was hard to believe. It was an exclusive storehouse for dental supplies and equipment. Paul told me the shipments kept arriving, and obviously they would never use it all. Well, you can imagine what evolved. When he heard about our shortages, he told me to take what I needed. His assurance that no one would miss or even realize the reduction in supplies was all I needed. This was the beginning of my periodic visits to Forbes, where I loaded my car with everything basic plus the newest technical do-dads. Back at Riley, I made the rounds distributing the rare goodies. My standard reply to questions (especially from my Colonel—an old time dentist whose acumen seemed to have plateaued somewhere around WW1) about where I got the stuff, was to say,

" It was best not to know, but you could think of it as a gift from on high. "

~ * ~

Though it is comforting to recognize that nuclear war hasn't materialized, its possibility can't be dismissed. While pursuing a hopeful future, the specter of this sobering reality lurks deep in our consciousness. I for one, place this concern among the myriad forces in this world which I can't influence. So I simply forget about it, and focus my energy on those choices and activities I can deal with. Of course, fixing teeth met this objective as an exquisitely tangible occupation, in which I have productively spent much of my life.

-END-

www.ingramcontent.com/pod-product-compliance
Lightning Source LLC
Chambersburg PA
CBHW071223290326
41931CB00037B/1953